Dedication

To the memory of my nanny,
Ms. Kwai-Sheung Ng (1893–1986)

The Author

Deborah D. L. Chung is Niagara Mohawk Power Corporation Endowed Chair Professor, Director of the Composite Materials Research Laboratory, and Professor of Mechanical and Aerospace Engineering at the State University of New York (SUNY) in Buffalo. She holds a Ph.D. in materials science and an S.M. degree from the Massachusetts Institute of Technology (M.I.T.), as well as an M.S. in engineering science and a B.S. in engineering and applied science from the California Institute of Technology.

Dr. Chung is a Fellow of ASM International and of the American Carbon Society, and is past recipient of the Teacher of the Year Award from Tau Beta Pi; the Teetor Educational Award from the Society of Automotive Engineers; the Hardy Gold Medal from the American Institute of Mining, Metallurgical, and Petroleum Engineers; and the Ladd Award from Carnegie Mellon University.

Dr. Chung has written or cowritten 322 articles published in journals (88 on carbon, 107 on cement-matrix composites, 31 on metal-matrix composites, 62 on polymer-matrix composites, 12 on metal-semiconductor interfaces, 5 on silicon, and 17 on other topics). She is the author of three books, including *Carbon Fiber Composites* (Butterworth, 1994) and *Composite Materials for Electronic Functions* (Trans Tech, 2000), and has edited two books including *Materials for Electronic Packaging* (Butterworth, 1995).

Dr. Chung is the holder of 16 patents and has given 125 invited lectures. Her research has covered many materials, including lightweight structural, construction, smart, adsorption, battery electrode, solar cell, and electronic packaging materials.

Preface

Materials constitute the foundation of technology. They include metals, polymers, ceramics, semiconductors, and composite materials. The fundamental concepts of materials science are crystal structures, imperfections, phase diagrams, materials processing, and materials properties. They are taught in most universities to materials, mechanical, aerospace, electrical, chemical, and civil engineering undergraduate students. However, students need to know not only the fundamental concepts, but also how materials are applied in the real world. Since a large proportion of undergraduate students in engineering go on to become engineers in various industries, it is important for them to learn about applied materials science.

Due to the multifunctionality of many materials and the breadth of industrial needs, this book covers structural, electronic, thermal, electrochemical, and other applications of materials in a cross-disciplinary fashion. The materials include metals, ceramics, polymers, cement, carbon, and composites. The topics are scientifically rich and technologically relevant. Each is covered in a tutorial and up-to-date manner with numerous references cited. The book is suitable for use as a textbook for undergraduate and graduate courses, or as a reference book. The reader should have background in fundamental materials science (at least one course), although some fundamental concepts pertinent to the topics in the chapters are covered in the appendices.

Contents

1 Introduction to Materials Applications

CONTENTS

SYNOPSIS Engineering materials constitute the foundation of technology, whether the technology pertains to structural, electronic, thermal, electrochemical, environmental, biomedical, or other applications. The history of human civilization evolved from the Stone Age to the Bronze Age, the Iron Age, the Steel Age, and to the Space Age (contemporaneous with the Electronic Age). Each age is marked by the advent of certain materials. The Iron Age brought tools and utensils. The Steel Age brought rails and the Industrial Revolution. The Space Age brought structural materials (e.g., composite materials) that are both strong and lightweight. The Electronic Age brought semiconductors. Modern materials include metals, polymers, ceramics, semiconductors, and composite materials. This chapter provides an overview of the classes and applications of materials.

RELEVANT APPENDICES: *A, B*

1.1 CLASSES OF MATERIALS

Metals, polymers, ceramics, semiconductors, and composite materials constitute the main classes of materials.

Metals (including alloys) consist of atoms and are characterized by metallic bonding (i.e., the valence electrons of each atom are delocalized and shared among

1

all the atoms). Most of the elements in the Periodic Table are metals. Examples of alloys are Cu-Zn (brass), Fe-C (steel), and Sn-Pb (solder). Alloys are classified according to the majority element present. The main classes of alloys are iron-based alloys for structures; copper-based alloys for piping, utensils, thermal conduction, electrical conduction, etc.; and aluminum-based alloys for lightweight structures and metal-matrix composites. Alloys are almost always in the polycrystalline form.

Ceramics are inorganic compounds such as Al_2O_3 (for spark plugs and for substrates for microelectronics), SiO_2 (for electrical insulation in microelectronics), Fe_3O_4 (ferrite for magnetic memories used in computers), silicates (clay, cement, glass, etc.), and SiC (an abrasive). The main classes of ceramics are oxides, carbides, nitrides, and silicates. Ceramics are typically partly crystalline and partly amorphous. They consist of ions (often atoms as well) and are characterized by ionic bonding and often covalent bonding.

Polymers in the form of thermoplastics (nylon, polyethylene, polyvinyl chloride, rubber, etc.) consist of molecules that have covalent bonding within each molecule and van der Waals' forces between them. Polymers in the form of thermosets (e.g., epoxy, phenolics, etc.) consist of a network of covalent bonds. Polymers are amorphous, except for a minority of thermoplastics. Due to the bonding, polymers are typically electrical and thermal insulators. However, conducting polymers can be obtained by doping, and conducting polymer-matrix composites can be obtained by the use of conducting fillers.

Semiconductors have the highest occupied energy band (the valence band, where the valence electrons reside energetically) full such that the energy gap between the top of the valence band and the bottom of the empty energy band (the conduction band) is small enough for some fraction of the valence electrons to be excited from the valence band to the conduction band by thermal, optical, or other forms of energy. Conventional semiconductors, such as silicon, germanium, and gallium arsenide (GaAs, a compound semiconductor), are covalent network solids. They are usually doped in order to enhance electrical conductivity. They are used in the form of single crystals without dislocations because grain boundaries and dislocations would degrade electrical behavior.

Composite materials are multiphase materials obtained by artificial combination of different materials to attain properties that the individual components cannot attain. An example is a lightweight structural composite obtained by embedding continuous carbon fibers in one or more orientations in a polymer matrix. The fibers provide the strength and stiffness while the polymer serves as the binder. Another example is concrete, a structural composite obtained by combining cement (the matrix, i.e., the binder, obtained by a reaction known as hydration, between cement and water), sand (fine aggregate), gravel (coarse aggregate), and, optionally, other ingredients known as admixtures. Short fibers and silica fume (a fine SiO_2 particulate) are examples of admixtures. In general, composites are classified according to their matrix materials. The main classes of composites are polymer-matrix, cement-matrix, metal-matrix, carbon-matrix, and ceramic-matrix.

Polymer-matrix and cement-matrix composites are the most common due to the low cost of fabrication. Polymer-matrix composites are used for lightweight structures (aircraft, sporting goods, wheelchairs, etc.) in addition to vibration damping, electronic enclosures, asphalt (composite with pitch, a polymer, as the matrix), and solder replacement. Cement-matrix composites in the form of concrete (with fine and coarse aggregates), steel-reinforced concrete, mortar (with fine aggregate, but no coarse aggregate), or cement paste (without any aggregate) are used for civil structures, prefabricated housing, architectural precasts, masonry, landfill cover, thermal insulation, and sound absorption. Carbon-matrix composites are important for lightweight structures (like the Space Shuttle) and components (such as aircraft brakes) that need to withstand high temperatures, but they are relatively expensive because of the high cost of fabrication. Carbon-matrix composites suffer from their tendency to be oxidized ($2C + O_2 \rightarrow 2CO$), thereby becoming vapor. Ceramic-matrix composites are superior to carbon-matrix composites in oxidation resistance, but they are not as well developed. Metal-matrix composites with aluminum as the matrix are used for lightweight structures and low-thermal-expansion electronic enclosures, but their applications are limited by the high cost of fabrication and by galvanic corrosion.

Not included in the five categories above is carbon, which can be in the common form of graphite, diamond, or fullerene (a recently discovered form). They are not ceramics because they are not compounds.

Graphite, a semimetal, consists of carbon atom layers stacked in the AB sequence such that the bonding is covalent due to sp^2 hybridization and metallic (two-dimensionally delocalized $2p_z$ electrons) within a layer, and is van der Waals between the layers. This bonding makes graphite very anisotropic, so it is a good lubricant due to the ease of the sliding of the layers with respect to one another. Graphite is also used for pencils because of this property. Moreover, graphite is an electrical and thermal conductor within the layers, but an insulator in the direction perpendicular to the layers. The electrical conductivity is valuable in its use for electrochemical electrodes. Graphite is chemically quite inert; however, due to anisotropy, it can undergo a reaction (known as intercalation) in which a foreign species called the intercalate is inserted between the carbon layers.

Disordered carbon (called turbostratic carbon) also has a layered structure, but, unlike graphite, it does not have the AB stacking order and the layers are bent. Upon heating, disordered carbon becomes more ordered, as the ordered form (graphite) has the lowest energy. Graphitization refers to the ordering process that leads to graphite. Conventional carbon fibers are mostly disordered carbon such that the carbon layers are along the fiber axis. Flexible graphite is formed by compressing a collection of intercalated graphite flakes that have been exfoliated (allowed to expand over 100 times along the direction perpendicular to the layers, typically through heating after intercalation). The exfoliated flakes are held together by mechanical interlocking because there is no binder. Flexible graphite is typically in the form of sheets, which are resilient in the direction perpendicular to the sheets. This resilience allows flexible graphite to be used as gaskets for fluid sealing.

Diamond is a covalent network solid exhibiting the diamond crystal structure due to sp^3 hybridization (akin to silicon). It is used as an abrasive and as a thermal conductor. Its thermal conductivity is the highest among all materials; however, it is an electrical insulator. Due to its high material cost, diamond is typically used in the form of powder or thin-film coating. Diamond is to be distinguished from diamond-like carbon (DLC), which is amorphous carbon that is sp^3-hybridized. Diamond-like carbon is mechanically weaker than diamond, but it is less expensive.

Fullerenes are molecules (C_{60}) with covalent bonding within each molecule. Adjacent molecules are held by van der Waals' forces; however, fullerenes are not polymers. Carbon nanotubes are derivatives of the fullerenes, as they are essentially fullerenes with extra carbon atoms at the equator. The extra atoms cause the fullerenes to be longer. For example, ten extra atoms (one equatorial band of atoms) exist in the molecule C_{70}. Carbon nanotubes can be single-wall or multiwall, depending on the number of carbon layers.

1.2 STRUCTURAL APPLICATIONS

Structural applications are applications that require mechanical performance (strength, stiffness, and vibration damping ability) in the material, which may or may not bear the load in the structure. In case the material bears the load, the mechanical property requirements are particularly exacting. An example is a building in which steel-reinforced concrete columns bear the load of the structure and unreinforced concrete architectural panels cover the face of the building. Both the columns and the panels serve structural applications and are structural materials, though only the columns bear the load. Mechanical strength and stiffness are required of the panels, but the requirements are more stringent for the columns.

Structures include buildings, bridges, piers, highways, landfill cover, aircraft, automobiles (body, bumper, drive shaft, window, engine components, and brakes), bicycles, wheelchairs, ships, submarines, machinery, satellites, missiles, tennis rackets, fishing rods, skis, pressure vessels, cargo containers, furniture, pipelines, utility poles, armored vehicles, utensils, fasteners, etc.

In addition to mechanical properties, a structural material may be required to have other properties, such as low density (lightweight) for fuel saving in the case of aircraft and automobiles, for high speed in the case of racing bicycles, and for handleability in the case of wheelchairs and armored vehicles. Another property often required is corrosion resistance, which is desirable for the durability of all structures, particularly automobiles and bridges. Yet another property that may be required is the ability to withstand high temperatures and/or thermal cycling, as heat may be encountered by the structure during operation, maintenance, or repair.

A relatively new trend is for a structural material to be able to serve functions other than the structural function. The material becomes multifunctional, thereby lowering cost and simplifying design. An example of a nonstructural function is the sensing of damage. Such sensing, also called structural health monitoring, is valuable for the prevention of hazards. It is particularly important to aging aircraft and bridges. The sensing function can be attained by embedding sensors (such as optical fibers, the damage or strain of which affects the light throughput) in the structure. However, embedding usually causes degradation of the mechanical properties, and the embedded devices are costly and poor in durability compared to the structural material. Another way to attain the sensing function is to detect the change in property (e.g., the electrical resistivity) of the structural material due to damage. In this way, the structural material serves as its own sensor and is said to be "self-sensing."

Mechanical performance is basic to the selection of a structural material. Desirable properties are high strength, high modulus (stiffness), high ductility, high toughness (energy absorbed in fracture), and high capacity for vibration damping. Strength, modulus, and ductility can be measured under tension, compression, or flexure at various loading rates as dictated by the type of loading on the structure. A high compressive strength does not imply a high tensile strength. Brittle materials tend to be stronger under compression than under tension because of microcracks. High modulus does not imply high strength, as modulus describes elastic deformation behavior, whereas strength describes fracture behavior. Low toughness does not imply a low capacity for vibration damping, as damping (energy dissipation) may be due to slipping at interfaces in the material rather than the shear of a viscoelastic phase. Other desirable mechanical properties are fatigue resistance, creep resistance, wear resistance, and scratch resistance.

Structural materials are predominantly metal-based, cement-based, and polymer-based, although they also include carbon-based and ceramic-based materials, which are valuable for high-temperature structures. Among the metal-based structural materials, steel and aluminum alloys are dominant. Steel is advantageous in high strength, whereas aluminum is advantageous in low density. For high-temperature applications, intermetallic compounds (such as NiAl) have emerged, though their brittleness is a disadvantage. Metal-matrix composites are superior to the corresponding metal matrices in high modulus, high creep resistance, and low thermal expansion coefficient, but they are expensive due to the processing cost.

Among the cement-based structural materials, concrete is dominant. Although concrete is an old material, improvement in long-term durability is needed, as suggested by the degradation of bridges and highways across the U.S. The improvement pertains to the decrease in drying shrinkage (shrinkage of the concrete during curing or hydration), as shrinkage can cause cracks. It also relates to the decrease in fluid permeability because water permeating into steel-reinforced concrete can cause corrosion of the reinforcing steel. Another area of improvement is freeze-thaw durability, which is the ability of the concrete to withstand temperature variations between 0°C and below (the freezing of water in concrete) and those above 0°C (the thawing of water in concrete).

Among polymer-based structural materials, fiber-reinforced polymers are dominant due to their combination of high strength and low density. All polymer-based materials suffer from the inability to withstand high temperatures. This inability may be due to the degradation of the polymer itself or, in the case of a polymer-matrix composite, thermal stress resulting from the thermal expansion mismatch between the polymer matrix and the fibers. (The coefficient of thermal expansion is typically much lower for the fibers than for the matrix.)

Most structures involve joints, which may be formed by welding, brazing, soldering, the use of adhesives, or by fastening. The structural integrity of joints is critical to the integrity of the overall structure.

As structures can degrade or be damaged, repair may be needed. Repair often involves a repair material, which may be the same as or different from the original material. For example, a damaged concrete column may be repaired by removing the damaged portion and patching with a fresh concrete mix. A superior but much more costly way involves the abovementioned patching, followed by wrapping the column with continuous carbon or glass fibers and using epoxy as the adhesive between the fibers and the column. Due to the tendency of the molecules of a thermoplastic polymer to move upon heating, the joining of two thermoplastic parts can be attained by liquid-state or solid-state welding. In contrast, the molecules of a thermosetting polymer do not move, so repair of a thermoset structure needs to involve other methods, such as adhesives.

Corrosion resistance is desirable for all structures. Metals, due to their electrical conductivity, are particularly prone to corrosion. In contrast, polymers and ceramics, because of their poor conductivity, are much less prone to corrosion. Techniques of corrosion protection include the use of a sacrificial anode (a material that is more active than the material to be protected, so that it is the part that corrodes) and cathodic protection (the application of a voltage that causes electrons to go into the material to be protected, thereby making the material a cathode). The first technique involves attaching the sacrificial anode material to the material to be protected. The second technique involves applying an electrical contact material on the surface of the material to be protected and passing an electric current through wires embedded in the electrical contact. The electrical contact material must be a good conductor and must be able to adhere to the material to be protected. It must also be wear resistant and scratch resistant.

Vibration damping is desirable for most structures. It is commonly attained by attaching to or embedding in the structure a viscoelastic material, such as rubber. Upon vibration, shear deformation of the viscoelastic material causes energy dissipation. However, due to the low strength and modulus of the viscoelastic material compared to the structural material, the presence of the viscoelastic material (especially if it is embedded) lowers the strength and modulus of the structure. A better way to attain vibration damping is to modify the structural material itself, so that it maintains its high strength and modulus while providing damping. If a composite material is the structural material, the modification can involve the addition of a filler (particles or fibers) with a very small size, so that the total filler-matrix interface area is large and slippage at the interface during vibration provides a mechanism of energy dissipation.

1.3 ELECTRONIC APPLICATIONS

Electronic applications include electrical, optical, and magnetic applications, as the electrical, optical, and magnetic properties of materials are largely governed by electrons. There is overlap among these three areas of application.

Electrical applications pertain to computers, electronics, electrical circuitry (resistors, capacitors, and inductors), electronic devices (diodes and transistors), optoelectronic devices (solar cells, light sensors, and light-emitting diodes for conversion between electrical energy and optical energy), thermoelectric devices (heaters, coolers, and thermocouples for conversion between electrical energy and thermal energy), piezoelectric devices (strain sensors and actuators for conversion between electrical energy and mechanical energy), robotics, micromachines (or microelectromechanical systems, MEMS), ferroelectric computer memories, electrical interconnections (solder joints, thick-film conductors, and thin-film conductors), dielectrics (electrical insulators in bulk, thick-film, and thin-film forms), substrates for thick films and thin films, heat sinks, electromagnetic interference (EMI) shielding, cables, connectors, power supplies, electrical energy storage, motors, electrical contacts and brushes (sliding contacts), electrical power transmission, and eddy current inspection (the use of a magnetically induced electrical current to indicate flaws in a material).

Optical applications have to do with lasers, light sources, optical fibers (materials of low optical absorptivity for communication and sensing), absorbers, reflectors and transmitters of electromagnetic radiation of various wavelengths (for optical filters, low-observable or Stealth aircraft, radomes, transparencies, optical lenses, etc.), photography, photocopying, optical data storage, holography, and color control.

Magnetic applications relate to transformers, magnetic recording, magnetic computer memories, magnetic field sensors, magnetic shielding, magnetically levitated trains, robotics, micromachines, magnetic particle inspection (the use of magnetic particles to indicate the location of flaws in a magnetic material), magnetic energy storage, magnetostriction (strain in a material due to the application of a magnetic field), magnetorheological fluids (for vibration damping that is controlled by a magnetic field), magnetic resonance imaging (MRI, for patient diagnosis in hospitals), and mass spectrometry (for chemical analysis).

All classes of materials are used for electronic applications. Semiconductors are at the heart of electronic and optoelectronic devices. Metals are used for electrical interconnections, EMI shielding, cables, connectors, electrical contacts, and electrical power transmission. Polymers are used for dielectrics and cable jackets. Ceramics are used for capacitors, thermoelectric devices, piezoelectric devices, dielectrics, and optical fibers.

Microelectronics refers to electronics involving integrated circuits. Due to the availability of high-quality single crystalline semiconductors, the most critical problems the microelectronic industry faces do not pertain to semiconductors, but are related to electronic packaging, including chip carriers, electrical interconnections, dielectrics, heat sinks, etc. Section 3.2 provides more details on electronic packaging applications.

Because of the miniaturization and increasing power of microelectronics, heat dissipation is critical to performance and reliability. Materials for heat transfer from electronic packages are needed. Ceramics and polymers are both dielectrics, but ceramics are advantageous because of their higher thermal conductivity compared to polymers. Materials that are electrically insulating but thermally conducting are needed. Diamond is the best material that exhibits such properties, but it is expensive.

Because of the increasing speed of computers, signal propagation delay needs to be minimized by the use of dielectrics with low values of the relative dielectric constant. (A dielectric with a high value of the relative dielectric constant and that is used to separate two conductor lines acts like a capacitor, thereby causing signal propagation delay.) Polymers have the advantage over ceramics because of their low value of the relative dielectric constant.

Electronic materials are in the following forms: bulk (single crystalline, polycrystalline, or, less commonly, amorphous), thick film (typically over 10 μm thick, obtained by applying a paste on a substrate by screen printing such that the paste contains the relevant material in particle form, together with a binder and a vehicle), or thin film (typically less than 1500 Å thick, obtained by vacuum evaporation, sputtering, chemical vapor deposition, molecular beam epitaxy, or other techniques). Semiconductors are typically in bulk single-crystalline form (cut into slices called "wafers," each of which may be subdivided into "chips"), although bulk polycrystalline and amorphous forms are emerging for solar cells due to the importance of low cost. Conductor lines in microelectronics are mostly in thick-film and thin-film forms.

The dominant material for electrical connections is solder (Sn-Pb alloy). However, the difference in CTE between the two members that are joined by the solder causes the solder to experience thermal fatigue upon thermal cycling encountered during operation. Thermal fatigue can lead to failure of the solder joint. Polymer-matrix composites in paste form and containing electrically conducting fillers are being developed to replace solder. Another problem lies in the poisonous lead used in solder to improve the rheology of the liquid solder. Lead-free solders are being developed.

Heat sinks are materials with high thermal conductivity that are used to dissipate heat from electronics. Because they are joined to materials of a low CTE (e.g., a printed circuit board in the form of a continuous fiber polymer-matrix composite), they need to have a low CTE also. Hence, materials exhibiting both a high thermal conductivity and a low CTE are needed for heat sinks. Copper is a metal with a high thermal conductivity, but its CTE is too high. Therefore, copper is reinforced with continuous carbon fibers, molybdenum particles, or other fillers of low CTE.

1.4 THERMAL APPLICATIONS

Thermal applications are applications that involve heat transfer, whether by conduction, convection, or radiation. Heat transfer is needed in heating of buildings; in industrial processes such as casting and annealing, cooking, de-icing, etc., and in cooling of buildings, refrigeration of food and industrial materials, cooling of electronics, removal of heat generated by chemical reactions such as the hydration of cement, removal of heat generated by friction or abrasion as in a brake and as in machining, removal of heat generated by the impingement of electromagnetic radiation, removal of heat from industrial processes such as welding, etc.

Conduction refers to the heat flow from points of higher temperature to points of lower temperature in a material. It typically involves metals because of their high thermal conductivity.

Convection is attained by the movement of a hot fluid. If the fluid is forced to move by a pump or a blower, the convection is known as forced convection. If the fluid moves due to differences in density, the convection is known as natural or free convection. The fluid can be a liquid (oil) or a gas (air) and must be able to withstand the heat involved. Fluids are outside the scope of this book.

Radiation, i.e., blackbody radiation, is involved in space heaters. It refers to the continual emission of radiant energy from the body. The energy is in the form of electromagnetic radiation, typically infrared. The dominant wavelength of the emitted radiation decreases with increasing temperature of the body. The higher the temperature, the greater the rate of emission of radiant energy per unit area of the surface. This rate is proportional to T^4, where T is the absolute temperature. It is also proportional to the emissivity of the body, and the emissivity depends on the material of the body. In particular, it increases with increasing roughness of the surface.

Heat transfer can be attained by the use of more than one mechanism. For example, both conduction and forced convection are involved when a fluid is forced to flow through the interconnected pores of a solid, which is a thermal conductor.

Conduction is tied more to material development than convection or radiation. Materials for thermal conduction are specifically addressed in Chapter 2.

Thermal conduction can involve electrons, ions, and/or phonons. Electrons and ions move from a point of higher temperature to a point of lower temperature, thereby transporting heat. Due to the high mass of ions compared to electrons, electrons move much more easily. Phonons are lattice vibrational waves, the propagation of which also leads to the transport of heat. Metals conduct by electrons because they have free electrons. Diamond conducts by phonons because free electrons are not available, and the low atomic weight of carbon intensifies the lattice vibrations. Diamond is the material with the highest thermal conductivity. In contrast, polymers are poor conductors because free electrons are not available and the weak secondary bonding (van der Waals' forces) between the molecules makes it difficult for the phonons to move from one molecule to another. Ceramics tend to be more conductive than polymers due to ionic and covalent bonding, making it possible for the phonons to propagate. Moreover, ceramics tend to have more mobile electrons or ions than polymers, and the movement of electrons and/or ions contributes to thermal conduction. On the other hand, ceramics tend to be poorer than metals in thermal conductivity because of the low concentration of free electrons (if any) in ceramics compared to metals.

1.5 ELECTROCHEMICAL APPLICATIONS

Electrochemical applications are applications that pertain to electrochemical reactions. An electrochemical reaction involves an oxidation reaction (such as $Fe \rightarrow Fe^{2+} + 2e^-$) in which electrons are generated, and a reduction reaction (such as $O_2 + 2H_2O + 4e^- \rightarrow 4OH^-$) in which electrons are consumed. The electrode that releases electrons is the anode; the electrode that receives electrons is the cathode.

When the anode and cathode are electrically connected, electrons move from the anode to the cathode. Both the anode and cathode must be electronic conductors. As the electrons move in the wire from the anode to the cathode, ions move in an ionic conductor (called the electrolyte) placed between the anode and the cathode such that cations (positive ions) generated by the oxidation of the anode move in the electrolyte from the anode to the cathode.

Whether an electrode behaves as an anode or a cathode depends on its propensity for oxidation. The electrode that has the higher propensity serves as the anode, while the other electrode serves as the cathode. On the other hand, a voltage can be applied between the anode and the cathode at the location of the wire such that the positive end of the voltage is at the anode side. The positive end attracts electrons, thus forcing the anode to be oxidized, even when it may not be more prone to oxidation than the cathode.

The oxidation reaction is associated with corrosion of the anode. For example, the oxidation reaction $Fe \rightarrow Fe^{2+} + 2e^-$ causes iron atoms to be corroded away, becoming Fe^{2+} ions, which go into the electrolyte. The hindering of the oxidation reaction results in corrosion protection.

Electrochemical reactions are relevant not only to corrosion, but also to batteries, fuel cells, and industrial processes (such as the reduction of Al_2O_3 to make Al) that make use of electrochemical reactions. The burning of fossil fuels such as coal and gasoline causes pollution of the environment. In contrast, batteries and fuel cells cause fewer environmental problems.

A battery involves an anode and a cathode that are inherently different in their propensities for oxidation. When the anode and cathode are open-circuited at the wire, a voltage difference is present between them such that the negative end of the voltage is at the anode side. This is because the anode wants to release electrons, but the electrons cannot come out because of the open circuit condition. This voltage difference is the output of the battery, which is a source of direct current (DC).

A unit involving an anode and a cathode is called a "galvanic cell." A battery consists of a number of galvanic cells connected in series, so that the battery voltage is the sum of the voltages of the individual cells.

An example of a battery is the lead storage battery used in cars. Lead (Pb) is the anode, while lead dioxide (PbO_2, in the form of a coating on the lead) is the cathode. Sulfuric acid (H_2SO_4) is the electrolyte. The oxidation reaction (anode reaction) is

$$Pb + HSO_4^- \rightarrow PbSO_4 + H^+ + 2e^-$$

The reduction reaction (cathode reaction) is

$$PbO_2 + HSO_4^- + 3H^+ + 2e^- \rightarrow PbSO_4 + 2H_2O$$

Discharge is the state of operation of the battery. The $PbSO_4$ is a solid reaction product that adheres to the electrodes, hindering further reaction. A battery needs to be charged by forcing current through the battery in the opposite direction, thereby breaking down $PbSO_4$, i.e., making the above reactions go in the reverse direction. In a car, the battery is continuously charged by an alternator.

Another example of a battery is the alkaline version of the dry cell battery. This battery comprises a zinc anode and an MnO_2 cathode. Because MnO_2 is not an electrical conductor, carbon powder (an electrical conductor) is mixed with the MnO_2 powder in forming the cathode. The electrolyte is either KOH or NaOH. The anode reaction is

$$Zn + 2OH^- \rightarrow ZnO + H_2O + 2e^-$$

The cathode reaction is

$$2MnO_2 + H_2O + 2e^- \rightarrow Mn_2O_3 + 2OH^-$$

A fuel cell is a galvanic cell in which the reactants are continuously supplied. An example is the hydrogen-oxygen fuel cell. The anode reaction is

$$2H_2 + 4OH^- \rightarrow 4H_2O + 4e^-$$

The cathode reaction is

$$4e^- + O_2 + 2H_2O \rightarrow 4OH^-$$

The overall cell reaction (the anode and cathode reactions added together) is

$$2H_2 \; (g) + O_2 \; (g) \rightarrow 2H_2O \; (l)$$

which is the formation of water from the reaction of hydrogen and oxygen.

During cell operation, hydrogen gas is fed to a porous carbon plate that contains a catalyst that helps the anode reaction. The carbon is an electrical conductor, which allows electrons generated by the anode reaction to flow. The porous carbon is known as a "current collector." Simultaneously, oxygen gas is fed to another porous carbon plate that contains a catalyst. The two carbon plates are electrically connected by a wire; electrons generated by the anode reaction at one plate flow through the wire and enter the other carbon plate for consumption in the cathode reaction. As this occurs, the OH^- ions generated by the cathode reaction move through the electrolyte (KOH) between the two carbon plates, and then are consumed in the anode reaction at the other carbon plate. The overall cell reaction produces H_2O, which comes out of the cell at an opening located at the electrolyte between the two carbon plates. The useful output of the cell is the electric current associated with the flow of electrons in the wire from one plate to the other.

Materials required for electrochemical applications include the electrodes, current collector (such as the porous carbon plates of the fuel cell mentioned above), conductive additive (such as carbon powder mixed with the MnO_2 powder in a dry cell), and electrolyte. An electrolyte can be a liquid or a solid, as long as it is an ionic conductor. The interface between the electrolyte and an electrode is intimate and greatly affects cell performance. The ability to recharge a cell is governed by the reversibility of the cell reactions. In practice, the reversibility is not complete, leading to low charge-discharge cycle life.

1.6   ENVIRONMENTAL APPLICATIONS

Environmental applications are applications that pertain to protecting the environment from pollution. The protection can involve the removal of a pollutant or the reduction in the amount of pollutant generated. Pollutant removal can be attained by extraction through adsorption on the surface of a solid (e.g., activated carbon) with surface porosity. It can also be attained by planting trees, which take in CO_2 gas. Pollutant generation can be reduced by changing the materials and/or processes used in industry by using biodegradable materials (materials that can be degraded by Nature so that their disposal is not necessary), by using materials that can be recycled, or by changing the energy source from fossil fuels to batteries, fuel cells, solar cells, and/or hydrogen.

Materials have been developed mainly for structural, electronic, thermal, or other applications without much consideration of disposal or recycling problems. It is now recognized that such considerations must be included during the design and development of materials rather than after the materials have been developed.

Materials for adsorption are central to the development of materials for environmental applications. They include carbons, zeolites, aerogels, and other porous materials. Desirable qualities include large adsorption capacity, pore size large enough for relatively large molecules and ions to lodge in, ability to be regenerated or cleaned after use, fluid dynamics for fast movement of the fluid from which the pollutant is to be removed, and, in some cases, selective adsorption of certain species.

Activated carbon fibers are superior to activated carbon particles in fluid dynamics due to the channels between the fibers. However, they are much more expensive.

Pores on the surface of a material must be accessible from the outside in order to serve as adsorption sites. In general, the pores can be macropores (> 500 Å), mesopores (between 20 and 500 Å), micropores (between 8 and 20 Å), or micromicropores (less than 8 Å). Activated carbons typically have micropores and micromicropores.

Electronic pollution is an environmental problem that has begun to be important. It arises from the electromagnetic waves (particularly radio waves) that are present in the environment due to radiation sources such as cellular telephones. Such radiation can interfere with digital electronics such as computers, thereby causing hazards and affecting society's operation. To alleviate this problem, radiation sources and electronics are shielded by materials that reflect and/or absorb radiation. Chapter 4 addresses shielding materials.

1.7 BIOMEDICAL APPLICATIONS

Biomedical applications pertain to the diagnosis and treatment of conditions, diseases, and disabilities, as well as their prevention. They include implants (hips, heart valves, skin, and teeth), surgical and diagnostic devices, pacemakers (devices for electrical control of heartbeats), electrodes for collecting or sending electrical or optical signals for diagnosis or treatment, wheelchairs, devices for helping the disabled, exercise equipment, pharmaceutical packaging (for controlled release of a drug to the body or for other purposes), and instrumentation for diagnosis and chemical analysis (such as equipment for analyzing blood and urine). Implants are particularly challenging; they need to be made of materials that are biocompatible (compatible with fluids such as blood), corrosion resistant, wear resistant, and fatigue resistant, and must be able to maintain these properties over tens of years.

BIBLIOGRAPHY

Askeland, Donald R., *The Science and Engineeing of Materials*, 3rd Ed., PWS Pub. Co., Boston (1994).

Callister, William D., Jr., *Materials Science and Engineering: An Introduction*, 5th Ed., Wiley, New York (2000).

Chung, D.D.L., *Carbon Fiber Composites*, Butterworth-Heinemann, Boston (1994).

Chung, D.D.L., Ed., *Materials for Electronic Packaging*, Butterworth-Heinemann, Boston (1995).

Coletta, Vincent P., *College Physics*, Mosby, St. Louis (1995).

Ohring, Milton, *Engineering Materials Science*, Academic Press, San Diego (1995).

Schaffer, James P., Saxena, Ashok, Antolovich, Stephen D., Sanders, Thomas H., Jr., and Warner, Steven B., *The Science and Design of Engineering Materials*, 2nd Ed., McGraw-Hill, Boston (1995).

Shackelford, James F., *Introduction to Materials Science for Engineers*, 5th Ed., Prentice Hall, Upper Saddle River, NJ (2000).

Smith, William F., *Principles of Materials Science and Engineering*, 3rd Ed., McGraw-Hill, New York (1996).

White, Mary Anne, *Properties of Materials*, Oxford University Press, New York (1999).

Zumdahl, Steven S., *Chemistry*, 3rd Ed., D.C. Heath and Company, Lexington, MA (1993).

2 Materials for Thermal Conduction

SYNOPSIS Materials for thermal conduction are reviewed. They include materials exhibiting high thermal conductivity (such as metals, carbons, ceramics, and composites), and thermal interface materials (such as polymer-based and silicate-based pastes and solder).

2.1 INTRODUCTION

The transfer of heat by conduction is involved in the use of a heat sink to dissipate heat from an electronic package, the heating of an object on a hot plate, the operation of a heat exchanger, the melting of ice on an airport runway by resistance heating, the heating of a cooking pan on an electric range, and in numerous industrial processes that involve heating or cooling. Effective transfer of heat by conduction requires materials of high thermal conductivity. In addition, it requires a good thermal contact between the two surfaces in which heat transfer occurs. Without good thermal contacts, the use of expensive thermal conducting materials for the components is a waste. The attainment of a good thermal contact requires a thermal interface material,

TABLE 2.1
Thermal Properties and Density of Various Materials

Material	Thermal Conductivity (W/m.K)	Coefficient of Thermal Expansion (10^{-6} °C^{-1})	Density (g/cm^3)
Aluminum	247	23	2.7
Gold	315	14	19.32
Copper	398	17	8.9
Lead	30	39	11
Molybdenum	142	4.9	10.22
Tungsten	155	4.5	19.3
Invar	10	1.6	8.05
Kovar	17	5.1	8.36
Diamond	2000	0.9	3.51
Beryllium oxide	260	6	3
Aluminum nitride	320	4.5	3.3
Silicon carbide	270	3.7	3.3

such as a thermal grease, which must be thin between the mating surfaces, must conform to the topography of the mating surfaces, and should preferably have a high thermal conductivity. This chapter is a review of materials for thermal conduction, including materials of high thermal conductivity and thermal interface materials.

2.2 MATERIALS OF HIGH THERMAL CONDUCTIVITY

2.2.1 Metals, Diamond, and Ceramics

Table 2.1 provides the thermal conductivity of various metals. Copper is most commonly used when materials of high thermal conductivity are required. However, copper suffers from a high value of the coefficient of thermal expansion (CTE). A low CTE is needed when the adjoining component has a low CTE. When the CTEs of the two adjoining materials are sufficiently different and the temperature is varied, thermal stress occurs and may even cause warpage. This is the case when copper is used as a heat sink for a printed wiring board, which is a continuous fiber polymer-matrix composite that has a lower CTE than copper. Molybdenum and tungsten have low CTE, but their thermal conductivity is poor compared to copper.

The alloy Invar® (64Fe-36Ni) is outstandingly low in CTE among metals, but it is very poor in thermal conductivity. Diamond is most attractive, as it has very high thermal conductivity and low CTE, but it is expensive. Aluminum is not as conductive as copper, but it has a low density, which is attractive for aircraft electronics and applications (e.g., laptop computers) which require low weight.[2,3] Aluminum nitride is not as conductive as copper, but it is attractive in its low CTE. Diamond and most ceramic materials are very different from metals in their electrical insulation abilities. In contrast, metals are conducting both thermally and electrically.

For applications that require thermal conductivity and electrical insulation, diamond and appropriate ceramic materials can be used, but metals cannot.

2.2.2 METAL-MATRIX COMPOSITES

One way to lower the CTE of a metal is to form a metal-matrix composite[1] by using a low CTE filler. Ceramic particles such as AlN and SiC are used for this purpose because of their combination of high thermal conductivity and low CTE. As the filler usually has lower CTE and lower thermal conductivity than the metal matrix, the higher the filler volume fraction in the composite, the lower the CTE, and the lower the thermal conductivity.

Metal-matrix composites with discontinuous fillers are attractive for their processability into various shapes. However, layered composites in the form of a matrix-filler-matrix sandwich are useful for planar components. Discontinuous fillers are most commonly ceramic particles. The filler sheets are usually low-CTE metal alloy sheets (e.g., Invar® or 64Fe-36Ni, and Kovar® or 54Fe-29Ni-17Co). Aluminum and copper are common metal matrices because of their high conductivity.

2.2.2.1 Aluminum-Matrix Composites

Aluminum is the most dominant matrix for metal-matrix composites for both structural and electronic applications. This is because of its low cost and low melting point (660°C), facilitating composite fabrication by methods that involve melting.

Liquid-phase methods for the fabrication of metal-matrix composites include liquid metal infiltration, which usually involves using pressure from a piston or compressed gas to push the molten metal into the pores of a porous preform comprising the filler (particles that are not sintered) and a small amount of a binder.[4-6] Pressureless infiltration is less common, but is possible.[7,8] The binder prevents the filler particles from moving during infiltration, and also provides sufficient compressive strength to the preform so that it will not be deformed during infiltration. This method thus provides near-net-shape fabrication, i.e., the shape and size of the composite product are the same as those of the preform. Since machining of the composite is far more difficult than that of the preform, near-net-shape fabrication is desirable.

In addition to near-net-shape fabrication capability, liquid metal fabrication is advantageous in that it provides composites with high filler volume fractions (up to 70%). A high filler volume fraction is necessary to attain a low enough CTE ($< 10 \times 10^{-6}/°C$) in the composite, even if the filler is a low-CTE ceramic (e.g., SiC), since the aluminum matrix has a relatively high CTE.[9,10] However, to attain a high volume fraction using liquid metal infiltration, the binder used must be in a small amount (so as to not clog the pores in the preform) and still be effective. Hence, the binder technology[11-13] is critical.

The ductility of a composite decreases as the filler volume fraction increases, so a composite with a low enough CTE is quite brittle. Although the brittleness is not acceptable for structural applications, it is acceptable for electronic applications.

Another liquid-phase technique is stir casting,[1] which involves stirring the filler in the molten metal and then casting. This method suffers from the nonuniform distribution of the filler in the composite due to the difference in density between filler and molten metal, and the consequent tendency of the filler to either float or sink in the molten metal prior to solidification. Stir casting also suffers from the incapability of producing composites with high filler volume fractions.

Yet another liquid-phase technique is plasma spraying,[14] which involves spraying a mixture of molten metal and filler onto a substrate. This method suffers from the relatively high porosity of the resulting composite and the consequent need for densification by hot isostatic pressing or other methods, which are expensive.

A solid-phase technique is powder metallurgy, which involves mixing the matrix metal powder and filler and the subsequent sintering under heat and pressure.[14] This method is relatively difficult for the aluminum matrix because aluminum has a protective oxide, and the oxide layer on the surface of each aluminum particle hinders sintering. Furthermore, this method is usually limited to low-volume fractions of the filler.

The most common filler used is silicon carbide (SiC) particles due to the low cost and low CTE of SiC.[15] However, SiC suffers from its reactivity with aluminum. The reaction is

$$3SiC + 4Al \rightarrow 3Si + Al_4C_3$$

It becomes more severe as the composite is heated. The aluminum carbide is a brittle reaction product that lines the filler-matrix interface of the composite, thus weakening the interface. Silicon, the other reaction product, dissolves in the aluminum matrix, lowering the melting temperature of the matrix and causing nonuniformity in the phase distribution and mechanical property distribution.[16] Also, the reaction consumes a part of the SiC filler.[17]

A way to diminish this reaction is to use an Al-Si alloy matrix, since the silicon in the alloy matrix promotes the opposite reaction. However, the Al-Si matrix is less ductile than the Al matrix, causing the mechanical properties of the Al-Si matrix composite to be very poor compared to those of the corresponding Al-matrix composite. Therefore, the use of an Al-Si alloy matrix is not a solution to the problem.

An effective solution is to replace Si-C by aluminum nitride (AlN) particles, which do not react with aluminum, resulting in superior mechanical properties in the composite.[18] The fact that AlN has a higher thermal conductivity than SiC helps the thermal conductivity of the composite. Since the cost of the composite fabrication process dominates the cost of producing composites, the higher material cost of AlN compared to SiC does not matter, especially for electronic packaging. Aluminum oxide (Al_2O_3) also does not react with aluminum, but it is low in thermal conductivity and suffers from particle agglomeration.[18]

Other than ceramics such as SiC and AlN, a filler used in aluminum-matrix composites is carbon in the form of fibers of diameter around 10 μm[19-23] and, less commonly, filaments of diameter less than 1 μm.[24] Carbon also suffers from reactivity with aluminum to form aluminum carbide. However, fibers are more effective than particles for reducing the CTE of the composite. Carbon fibers can even be continuous in length. Moreover, carbon, especially when graphitized, is much more thermally conductive than ceramics. In fact, carbon fibers that are sufficiently graphitic are even more thermally conductive than the metal matrix, so the thermal conductivity of the composite increases with increasing fiber volume fraction. However, these fibers are expensive. The mesophase-pitch-based carbon fiber K-1100 from BP Amoco Chemicals exhibits longitudinal thermal conductivity of 1000 W/m.K.[25,26]

Both carbon and SiC form a galvanic couple with aluminum, which is the anode. It is the component in the composite that is corroded. The corrosion becomes more severe in the presence of heat and/or moisture.

The thermal conductivity of aluminum-matrix composites depends on the filler and its volume fraction, the alloy matrix heat treatment condition, and the filler-matrix interface.[18,27]

To increase the thermal conductivity of SiC aluminum-matrix composite, a diamond film can be deposited on the composite.[28] The thermal conductivity of single crystal diamond is 2000–2500 W/m.K, though a diamond film is not single crystalline.

2.2.2.2 Copper-Matrix Composites

Because copper is heavy, the filler does not have to be lightweight. Thus, low CTE but heavy metals such as tungsten,[29,30] molydenum,[31,32] and Invar®[33-35] are used as fillers. These metals (except Invar) have the advantage that they are quite conductive thermally and are available in particle and sheet forms; they are suitable for particulate as well as layered[36,37] composites. Another advantage of the metal fillers is better wettability of molten matrix metal with metal fillers than with ceramic fillers, which is important if the composite is fabricated by a liquid phase method.

An advantage of copper over aluminum is its nonreactivity with carbon, so carbon is a highly suitable filler for copper. Additional advantages are that carbon is lightweight and its fibers are available in a continuous form. Furthermore, copper is a rather noble metal, as shown by its position in the Electromotive Series, so it does not suffer from the corrosion that plagues aluminum. Carbon used as a filler in copper is in the form of fibers of diameter around 10 μm.[22,38-45] As carbon fibers that are sufficiently graphitic are even more thermally conductive than copper, the thermal conductivity of a copper-matrix composite can exceed that of copper. Less common fillers for copper are ceramics such as silicon carbide, titanium diboride (TiB$_2$), and alumina.[46-48]

The melting point of copper is much higher than that of aluminum, so the fabrication of copper-matrix composites is commonly done by powder metallurgy, although liquid metal infiltration is also used.[22,49,50] In the case of liquid metal infiltration, the metal matrix is often a copper alloy (e.g., Cu-Ag) chosen for reduced melting temperature and good castability.[50]

Powder metallurgy conventionally involves mixing the metal matrix powder and the filler, and subsequently pressing and then sintering under either heat or both heat and pressure. The problem with this method is that it is limited to low-volume fractions of the filler. In order to attain high-volume fractions, a less conventional method of powder metallurgy is recommended. This method involves coating the matrix metal on the filler units, followed by pressing and sintering.[32,46,51,52] The mixing of matrix metal powder with the coated filler is not necessary, although it can be done to decrease the filler volume fraction in the composite. The metal coating on the filler forces the distribution of matrix metal to be uniform, even when the metal volume fraction is low (i.e., when the filler volume fraction is high). With the conventional method, the matrix metal distribution is not uniform when the filler volume fraction is high, causing porosity and the presence of filler agglomerates, in each of which the filler units directly touch one another; this microstructure results in low thermal conductivity and poor mechanical properties.

Continuous carbon fiber copper-matrix composites can be made by coating the fibers with copper and then diffusion bonding (i.e., sintering).[38,40,44,53-55] This method is akin to the abovementioned less conventional method of powder metallurgy.

Less common fillers used in copper include diamond powder,[50,56] aluminosilicate fibers,[57] and Ni-Ti alloy rod.[58] The Ni-Ti alloy is attractive for its negative CTE of $-21 \times 10^{-6}/°C$. The coating of a carbon fiber copper-matrix composite with a diamond film has been undertaken to enhance thermal conductivity.[43]

2.2.2.3 Beryllium-Matrix Composites

Beryllium oxide (BeO) has a high thermal conductivity (Table 2.1). Beryllium-matrix BeO-platelet composites with 20-60 vol.% BeO exhibit low density (2.30 g/cm^3 at 40 vol.% BeO, compared to 2.9 g/cm^3 for Al/SiC at 40 vol.% SiC), high thermal conductivity (232 W/m.K at 40 vol.% BeO, compared to 130 W/m.K for Al/SiC at 40 vol.% SiC), low CTE ($7.5 \times 10^{-6}/°C$ at 40 vol.% BeO, compared to $12.1 \times 10^{-6}/°C$ at 40 vol.% SiC), and high modulus (317 GPa at 40 vol.% BeO, compared to 134 GPa for Al/SiC at 40 vol.% SiC).[59,60]

2.2.3 Carbon-Matrix Composites

Carbon is an attractive matrix for composites for thermal conduction because of its thermal conductivity and low CTE. Furthermore, it is corrosion resistant and light-weight. Yet another advantage of the carbon matrix is its compatibility with carbon fibers, in contrast to the common reactivity between a metal matrix and its fillers. Hence, carbon fibers are the dominant filler for carbon-matrix composites. Composites with both filler and matrix of carbon are called carbon-carbon composites.[61] Their primary applications are heat sinks,[62] thermal planes,[63] and substrates.[64] There is considerable competition between carbon-carbon composites and metal-matrix composites for the same applications.

The main drawback of carbon-matrix composites is their high cost of fabrication, which involves making a pitch-matrix or resin-matrix composite and subsequent carbonization of the pitch or resin by heating at 1000–1500°C in an inert atmosphere. After carbonization, the porosity is substantial in the carbon matrix, so pitch or resin is impregnated into the composite and carbonization is carried out again. Quite a few impregnation-carbonization cycles are needed to reduce the porosity to an acceptable level, resulting in high cost. Graphitization by heating at 2000–3000°C in an inert atmosphere may follow carbonization in order to increase thermal conductivity, which increases with the degree of graphitization. However, graphitization is an expensive step. Some or all of the impregnation-carbonization cycles may be replaced by chemical vapor infiltration (CVI), in which a carbonaceous gas infiltrates the composite and decomposes to form carbon.

Carbon-carbon composites have been made by using conventional carbon fibers of diameters around 10 μm.[62,63,65] and by using carbon filaments grown catalytically from carbonaceous gases and of diameters less than 1 μm.[24] By using graphitized carbon fibers, thermal conductivities exceeding that of copper can be reached.

To increase thermal conductivity, carbon-carbon composites have been impregnated with copper[66,67] and have been coated with a diamond film.[68]

2.2.4 Carbon and Graphite

An all-carbon material (called ThermalGraph®, a tradename of BP Amoco Chemicals), made by consolidating oriented precursor carbon fibers without a binder and subsequent carbonization and optional graphitization, exhibits thermal conductivity ranging from 390 to 750 W/m.K in the fiber direction of the material.

Another material is pyrolytic graphite (called TPG) encased in a structural shell.[69] The graphite, highly textured with the c-axes of the grains perpendicular to the plane of the graphite, has an in-plane thermal conductivity of 1700 W/m.K (four times that of copper), but it is mechanically weak because of the tendency to shear in the plane of the graphite. The structural shell serves to strengthen by hindering shear.

Pitch-derived carbon foams, with thermal conductivity up to 150 W/m.K after graphitization, are attractive for their high specific thermal conductivity (thermal conductivity divided by the density).[70]

2.2.5 Ceramic-Matrix Composites

The SiC matrix is attractive due to its high CTE compared to the carbon matrix, though it is not as thermally conductive as carbon. The CTE of carbon-carbon composites is too low (0.25×10^{-6}/°C), resulting in reduced fatigue life in chip-on-board (COB) applications with silica chips (CTE = 2.6×10^{-6}/°C). The SiC-matrix carbon fiber composite is made from a carbon-carbon composite by converting the matrix from carbon to SiC.[65] To improve the thermal conductivity of the SiC-matrix composite, coatings in the form of chemical vapor-deposited AlN or Si have been used. The SiC-matrix metal (Al or Al-Si) composite, as made by a liquid-exchange process, also exhibits relatively high thermal conductivity.[71]

Borosilicate glass matrix is attractive due to its low dielectric constant (4.1 at 1 MHz for B_2O_3-SiO_2-Al_2O_3-Na_2O glass) compared to 8.9 for AlN, 9.4 for alumina (90%), 42 for SiC, 6.8 for BeO, 7.1 for cubic boron nitride, 5.6 for diamond, and 5.0 for glass-ceramic. A low value of the dielectric constant is desirable for electronic packaging applications. On the other hand, glass has a low thermal conductivity, so fillers with relatively high thermal conductivity are used with the glass matrix. An example is continuous SiC fibers, the glass-matrix composites of which are made by tape casting followed by sintering.[72] Another example is aluminum nitride with interconnected pores (about 28 vol.%), the composites of which are obtained by glass infiltration to a depth of about 100 μm.[72-74]

2.3 THERMAL INTERFACE MATERIALS

The improvement of a thermal contact involves the use of a thermal interface material, such as a thermal fluid, a thermal grease (paste), a resilient thermal conductor, or solder that is applied in the molten state. A thermal fluid, thermal grease, or molten solder is spread on the mating surfaces. A resilient thermal conductor is sandwiched by the mating surfaces and held in place by pressure. Thermal fluids are most commonly mineral oil. Thermal greases (pastes) are usually conducting particle-filled silicone. Resilient thermal conductors are conducting particle-filled elastomers. Of these four types of thermal interface materials, thermal greases (based on polymers, particularly silicone) and solder are by far the most common. Resilient thermal conductors are not as well developed as thermal fluids or greases.

As the materials to be interfaced are good thermal conductors, the effectiveness of a thermal interface material is enhanced by high thermal conductivity and low thickness of the interface material, and low thermal contact resistance between the interface material and each mating surface. As the mating surfaces are not perfectly smooth, the interface material must be able to flow or deform so as to conform to the topography of the mating surfaces. If the interface material is a fluid, grease, or paste, it should have a high fluidity (workability) in order to conform and to have a small thickness after mating. On the other hand, the thermal conductivity of the grease or paste increases with increasing filler content, and this is accompanied by a decrease in workability. Without a filler, as in the case of an oil, thermal conductivity is poor. A thermal interface material in the form of a resilient thermal conductor sheet (e.g., felt consisting of conducting fibers held together without a binder, and a resilient polymer-matrix composite containing a thermally conducting filler) cannot be as thin or conformable as one in the form of a fluid, grease, or paste, so its effectiveness requires a very high thermal conductivity.

Solder is commonly used as a thermal interface material for enhancing the thermal contact between two surfaces. This is because it can melt at rather low temperatures and the molten solder can flow and spread itself thinly on the adjoining surfaces, resulting in high thermal contact conductance at the interface between the solder and each of the adjoining surfaces. Furthermore, solder in the metallic solid state is a good thermal conductor. In spite of its high thermal conductivity, the thickness of the solder greatly influences its effectiveness as a thermal interface material. When solder is used as a thermal interface material between copper surfaces, increasing the thickness of solder from 10 to 30 μm increases the heat transfer time by 25%. The effect is akin to replacing solder with an interface material that is 80% lower in thermal conductivity. It is also akin to decreasing the thermal contact conductance of the solder-copper interface by 70%.[75]

Thermal pastes are predominantly based on polymers, particularly silicone,[76-79] although thermal pastes based on sodium silicate have been reported to be superior in providing high thermal contact conductance.[80] The superiority of sodium-silicate-based pastes over silicone-based pastes is primarily due to the low viscosity of sodium silicate compared to silicone, and the importance of high fluidity in the paste so that it can conform to the topography of the surfaces it interfaces.

A particularly attractive thermal paste is based on polyethylene glycol (PEG, a polymer) of a low molecular weight (400 amu).[81] These pastes are superior to silicone-based pastes and are as good as sodium-silicate-based pastes because of the low viscosity of PEG and the contribution to thermal conduction of lithium ions (a dopant) in the paste. Compared to sodium-silicate-based pastes, PEG-based pastes are advantageous in their long-term compliance, in contrast to the long-term rigidity of sodium silicate. Compliance is attractive for decreasing thermal stress, which can cause thermal fatigue.

Table 2.2 presents the thermal contact conductance for different thermal interface materials. Included in the comparison are results obtained with the same testing method on silicone-based paste, sodium-silicate-based pastes, and solder.[80,81] PEG (i.e., A) gives much higher thermal contact conductance (11.0×10^4 W/m^2.°C) than silicone (3.08×10^4 W/m^2.°C), due to its relatively low viscosity, but the conductance is lower than that given by sodium silicate (14.1×10^4 W/m^2.°C) in spite of its low viscosity because of the molecular nature of PEG. The addition of the Li salt (1.5 wt.%) to PEG (i.e., to obtain C) raises the conductance from 11.0×10^4 to 12.3×10^4 W/m^2.°C, even though the viscosity is increased. The further addition of water and DMF (i.e., F) raises the conductance to 16.0×10^4 W/m^2.°C and decreases the viscosity. Thus, the addition of water and DMF is very influential, as water and DMF help the dissociation of the lithium salt. The further addition of BN particles (18.0 vol.%) (i.e., F_2) raises the conductance to 18.9×10^4 W/m^2.°C. The positive effect of BN is also shown by comparing the results of C and D (which are without water or DMF) and by comparing the results of A and B (which are without Li$^+$). In the absence of the lithium salt, water and DMF also help, though not greatly, as shown by comparing A and J. The viscosity increases with the lithium salt content, as shown by comparing J, E, F, G, H, and I.

Comparison of E, F, G, H, and I shows that the optimum lithium salt content for the highest conductance is 1.5 wt.%. That an intermediate lithium salt content gives the highest conductance is probably because of the enhancement of the thermal conductivity by the Li$^+$ ions and the increase of the viscosity. Both high conductivity and low viscosity are desirable for a high-contact conductance. Comparison of F_1, F_2, F_3, and F_4 shows that the optimum BN content is 18.0 vol.%, as also indicated by comparing G_1, G_2, G_3, and G_4. Among all the PEG-based pastes, the highest conductance is given by F_2, as it has the optimum lithium salt content as well as the optimum BN content. An optimum BN content also occurs for BN-filled sodium-silicate-based pastes.[80] It is due to the increase in both thermal conductivity and viscosity as the BN content increases. The best PEG-based paste (i.e., F_2) is similar to the best sodium-silicate-based paste in conductance. Both are better than BN-filled silicone, but both are slightly inferior to solder. Although solder gives the highest conductance, it suffers from the need for heating during soldering. In contrast, heating is not needed for the use of PEG-based, silicone-based, or silicate-based pastes.

TABLE 2.2

Thermal Contact Conductance for Various Thermal Interface Materials between Copper Disks at 0.46 Ma Contact Pressure[80,81]

Thermal Interface Material		Interface Material Thickness (μm) (±10)	Thermal Contact Conductance (10^4 W/m^2.°C)	Viscosity (cps) (± 0.3)
Description	Designation			
PEG	A	<25	11.0 ± 0.3	127[b]
PEG + BN (18 vol.%)	B	25	12.3 ± 0.3	/
PEG + Li salt (1.5 wt.%)	C	<25	12.3 ± 0.3	143[b]
PEG + Li salt (1.5 wt.%) + BN (18 vol.%)	D	25	13.4 ± 0.4	/
PEG + water + DMF	J	<25	12.5 ± 0.2	75.6[b]
J + Li salt (0.75 wt.%)	E	<25	11.4 ± 0.3	79.7[b]
J + Li salt (1.5 wt.%)	F	<25	16.0 ± 0.5	85.6[b]
J + Li salt (3.0 wt.%)	G	<25	11.6 ± 0.2	99.0[b]
J + Li salt (4.5 wt. %)	H	<25	9.52 ± 0.25	117[b]
J + Li salt (6.0 wt.%)	I	<25	7.98 ± 0.16	120[b]
F + BN (16.0 vol.%)	F_1	25	18.5 ± 0.8	/
F + BN (18.0 vol.%)	F_2	25	18.9 ± 0.8	/
F + BN (19.5 vol. %)	F_3	25	15.3 ± 0.2	/
F + BN (21.5 vol.%)	F_4	25	14.0 ± 0.5	/
G + BN (16.0 vol.%)	G_1	25	17.0 ± 0.5	/
G + BN (18.0 vol.%)	G_2	25	17.3 ± 0.6	/
G + BN (19.5 vol.%)	G_3	25	14.9 ± 0.6	/
G + BN (21.5 vol.%)	G_4	25	13.4 ± 0.4	/
H + BN (18.0 vol.%)	H_1	25	13.9 ± 0.4	/
Solder	/	25	20.8 ± 0.6[a]	/
Sodium silicate + BN (16.0 vol. %)	/	25	18.2 ± 0.7	/
Sodium silicate + BN (17.3 vol. %)	/	25	15.5 ± 0.4	/
Sodium silicate	/	<25	14.1 ± 0.5	206[c]
Silicone/BN	/	25	10.9 ± 1.5	/
Silicone	/	<25	3.08 ± 0.03	8800[d]
None	/	/	0.681 ± 0.010	/

[a] At zero contact pressure (not 0.46 MPa)
[b] Measured using the Ubbelohde method
[c] Measured using the Ostwald method
[d] Value provided by the manufacturer

2.4 CONCLUSION

Materials for thermal conduction include those exhibiting high thermal conductivity, as well as thermal interface materials. The former includes metals, diamond, carbon, graphite, ceramics, metal-matrix composites, carbon-matrix composites, and ceramic-matrix composites. The latter includes polymer-based pastes, silicate-based pastes, and solder.

REFERENCES

1. P.K. Rohatgi, *Defence Sci. J.*, 43(4), 323-349 (1993).
2. W.M. Peck, American Society of Mechanical Engineers, Heat Transfer Division, HTD, 329(7), 245-253 (1996).
3. A.L. Geiger and M. Jackson, *Adv. Mater. Proc.*, 136(1), 6 (1989).
4. S. Lai and D.D.L. Chung, *J. Mater. Sci.*, 29, 3128-3150 (1994).
5. X. Yu, G. Zhang and R. Wu, *J. Mater. Eng.*, 6, 9-12 (June 1994).
6. B.E. Novich and R.W. Adams, Proc. 1995 Int. Electronics Packaging Conf., Int. Electronics Packaging Society, pp. 220-227 (1995).
7. X.F. Yang and X.M. Xi, *J. Mater. Res.*, 10(10), 2415-2417 (1995).
8. J.A. Hornor and G.E. Hannon, 6th Int. Electronics Packaging Conf., pp. 295-307 (1992).
9. Y.-L. Shen, A. Needleman and S. Suresh, *Met. Mater. Trans.*, 5A(4), 839-850 (1994).
10. Y.-L. Shen, *Mater. Sci. Eng. A*, 2, 269-275 (Sept. 1998).
11. J. Chiou and D.D.L. Chung, *J. Mater. Sci.*, 28, 1435-1446 (1993).
12. J. Chiou and D.D.L. Chung, *J. Mater. Sci.*, 28, 1447-1470 (1993).
13. J. Chiou and D.D.L. Chung, *J. Mater. Sci.*, 28, 1471-1487 (1993).
14. M.E. Smagorinski, P.G. Tsantrizos, S. Grenier, A. Cavasin, T. Brzezinski and G. Kim, *Mater. Sci. Eng. A*, 1, 86-90 (Mar. 1998).
15. K. Schmidt, C. Zweben and R. Arsenault, ASTM Special Technical Publication, No. 1080, pp. 155-164 (1990).
16. S. Lai and D.D.L. Chung, *J. Mater. Sci.*, 29, 2998-3016 (1994).
17. S. Lai and D.D.L. Chung, *J. Mater. Chem.*, 6, 469-477 (1996).
18. S. Lai and D.D.L. Chung, *J. Mater. Sci.*, 29, 6181-6198 (1994).
19. S. Wagner, M. Hook and E. Perkoski, Proc. Electricon '93, Electronics Manufacturing Productivity Facility, pp. 11/1-11/14 (1993).
20. O.J. Ilegbusi, *(Paper) ASME*, 93-WA/EEP-28, pp. 1-7 (1993).
21. M.A. Lambert and L.S. Fletcher, HTD Vol. 292, Heat Transfer in Electronic Systems, ASME, pp. 115-122 (1994).
22. M.A. Lambert and L.S. Fletcher, *J. Heat Transfer Trans.*, 118(2), 478-480 (1996).
23. M.A. Kinna, 6th Int. SAMPE Electronics Conf., pp. 547-555 (1992).
24. J.-M. Ting, M.L. Lake and D.R. Duffy, *J. Mater. Res.*, 10(6), 1478-1484 (1995).
25. T.F. Fleming, C.D. Levan and W.C. Riley, Proc. Technical Conf., Int. Electronics Packaging Conf., pp. 493-503 (1995).
26. T.F. Fleming and W.C. Riley, *Proc. SPIE — The Int. Soc. for Optical Eng.*, Vol. 1997, pp. 136-147 (1993).
27. H. Wang and S.H.J. Lo, *J. Mater. Sci. Lett.*, 15(5), 369-371 (1996).
28. R.E. Morgan, S.L. Ehlers and J. Sosniak, 6th Int. SAMPE Electronics Conf., pp. 320-333 (1992).
29. S. Yoo, M.S. Krupashankara, T.S. Sudarshan and R.J. Dowding, *Mater. Sci. Tech.*, 14(2), 170-174 (1998).
30. Anonymous, *Metal Powder Report*, 4, 28-31 (1997).
31. T.W. Kirk, S.G. Caldwell and J.J. Oakes, Particulate Materials and Processes, Advances in Powder Metallurgy, Metal Powder Industries Federation, Vol. 9, pp. 115-122 (1992).
32. P. Yih and D.D.L. Chung, *J. Electron. Mater.*, 24(7), 841-851 (1995).
33. S. Jha, 1995 Proc. 45th Electronic Components & Technology Conf., IEEE, pp. 542-547 (1995).
34. R. Chanchani and P.M. Hall, *IEEE Trans. Components Hybrids Manuf. Technol.*, 13(4), 743-750 (1990).

35. C. Woolger, *Materials World*, 4(6), 332-333 (1996).
36. R. Chanchani and P.M. Hall, Proc. 40th Electronic Components and Technology Conference, IEEE, pp. 94-102 (1990).
37. J.R. Hanson and J.L. Hauser, *Electron. Packaging Prod.*, 26(11), 48-51 (1986).
38. J. Korab, G. Korb and P. Sebo, Proc. IEEE/CPMT Int. Electronics Manufacturing Technology (IEMT) Symp., pp. 104-108 (1998).
39. J. Korab, G. Korb, P. Stefanik and H.P. Degischer, *Composites*, Part A, 30(8), 1023-1026 (1999).
40. G. Korb, J. Koráb and G. Groboth, *Composites*, Part A, 29A, 1563-1567 (1998).
41. M.A. Kinna, 6th Int. SAMPE Electronics Conf., pp., 547-555 (1992).
42. M.A. Lambert and L.S. Fletcher, HTD Vol. 292, Heat Transfer in Electronic Systems, ASME, pp. 115-122 (1994).
43. C.H. Stoessel, C. Pan, J.C. Withers, D. Wallace and R.O. Loutfy, *Mater. Res. Soc. Symp. Proc.*, Vol. 390, pp. 147-152 (1995).
44. Y. LePetitcorps, J.M. Poueylaud, L. Albingre, B. Berdeu, P. Lobstein and J.F. Silvain, *Key Engineering Materials*, 127-131, 327-334 (1997).
45. K. Prakasan, S. Palaniappan and S. Seshan, *Composites,* Part A, 28(12), 1019-1022 (1997).
46. P. Yih and D.D.L. Chung, *Int. J. Powder Met.*, 31(4), 335-340 (1995).
47. M. Ruhle, *Key Engineering Materials*, 116-117, 1-40 (1996).
48. M. Ruhle, *J. Eur. Ceramic Soc.*, 16(3), 353-365 (1996).
49. K. Prakasan, S. Palaniappan and S. Seshan, *Composites,* Part A, 28(12), 1019-1022 (1997).
50. J.A. Kerns, N.J. Colella, D. Makowiecki and H.L. Davidson, *Int. J. Microcircuits Electron. Packaging*, 19(3), 206-211 (1996).
51. P. Yih and D.D.L. Chung, *J. Mater. Sci.*, 32(7), 1703-1709 (1997).
52. P. Yih and D.D.L. Chung, *J. Mater. Sci.*, 32(11), 2873-2894 (1997).
53. C.H. Stoessel, J.C. Withers, C. Pan, D. Wallace and R.O. Loutfy, *Surf. Coatings Tech.*, 76-77, 640-644 (1995).
54. Y.Z. Wan, Y.L. Wang, G.J. Li, H.L. Luo and G.X. Cheng, *J. Mater. Sci. Lett.*, 16, 1561-1563 (1997).
55. D.A. Foster, 34th Int. SAMPE Symp. Exhib., Book 2, pp. 1401-1410 (1989).
56. H.L. Davidson, N.J. Colella, J.A. Kerns and D. Makowiecki, 1995 Proc. 45th Electronic Components & Technology Conf., IEEE, pp. 538-541 (1995).
57. K. Prakasan, S. Palaniappan and S. Seshan, *Composites,* Part A, 28(12), 1019-1022 (1997).
58. H. Mavoori and S. Jin, *JOM*, 50(6), 70-72 (1998).
59. T.B. Parsonage, 7th Int. SAMPE Electronics Conf., pp. 280-295 (1994).
60. T.B. Parsonage, National Electronic Packaging and Production Conf. — Proc. Technical Program (West and East), pp. 325-334 (1996).
61. K.M. Kearns, Proc. Int. SAMPE Symp. Exhib., 43(2), 1362-1369 (1998).
62. W.H. Pfeifer, J.A. Tallon, W.T. Shih, B.L. Tarasen and G.B. Engle, 6th Int. SAMPE Electronics Conf., pp. 734-747 (1992).
63. W.T. Shih, F.H. Ho and B.B. Burkett, 7th Int. SAMPE Electronics Conf., pp. 296-309 (1994).
64. W. Kowbel, X. Xia and J.C. Withers, 43rd Int. SAMPE Symp., pp. 517-527 (1998).
65. W. Kowbel, X. Xia, C. Bruce and J.C. Withers, *Mater. Res. Soc. Symp. Proc.*, Vol. 515, pp. 141-146 (1998).
66. S.K. Datta, S.M. Tewari, J.E. Gatica, W. Shih and L. Bentsen, *Met. Mater. Trans. A*, 30A, 175-181 (1999).

67. W.T. Shih, Proc. Int. Electronics Packaging Conf., pp. 211-219 (1995).

68. J.-M. Ting and M.L. Lake, *Diamond and Related Materials*, 3(10), 1243-1248 (1994.

69. M.J. Montesano, *Mat. Tech.,* 11(3), 87-91 (1996).

70. J. Klett, C. Walls and T. Burchell, Ext. Abstr. Program — 24th Bienn. Conf. Carbon, pp. 132-133 (1999).

71. L. Hozer, Y.-M. Chiang, S. Ivanova and I. Bar-On, *J. Mater. Res.*, 12(7), 1785-1789 (1997).

72. P.N. Kumta, *J. Mater. Sci.,* 31(23), 6229-6240 (1996).

73. P.N. Kumta, T. Mah, P.D. Jero and R.J. Kerans, *Mater. Lett.,* 21(3-4), 329-333 (1994).

74. J.Y. Kim and P.N. Kumta, Proc. IEEE 1998 National Aerospace and Electronics Conf., pp. 656-665 (1998).

75. X. Luo and D.D.L. Chung, *J. Electron. Packaging*, in press.

76. S.W. Wilson, A.W. Norris, E.B. Scott and M.R. Costello, National Electronic Packaging and Production Conference, Proc. Technical Program, Vol. 2, pp. 788-796 (1996).

77. A.L. Peterson, Proc. 40th Electronic Components and Technology Conf., IEEE, Vol. 1, pp. 613-619 (1990).

78. X. Lu, G. Xu, P.G. Hofstra and R.C. Bajcar, *J. Polymer Sci.,* 36(13), 2259-2265 (1998).

79. T. Sasaski, K. Hisano, T. Sakamoto, S. Monma, Y. Fijimori, H. Iwasaki and M. Ishizuka, Japan IEMT Symp. Proc. IEEE/CPMT Int. Electronic Manufacturing Technology (IEMT) Symp., pp. 236-239 (1995).

80. Y. Xu, X. Luo and D.D.L. Chung, *J. Electron. Packaging,* 122(2), 128-131 (2000).

81. Y. Xu, X. Luo and D.D.L. Chung, *J. Electron. Packaging,* in press.

3 Polymer-Matrix Composites for Microelectronics

CONTENTS

SYNOPSIS   Polymer-matrix composite materials for microelectronics are reviewed in terms of the science and applications. They include those with continuous and discontinuous fillers in the form of particles and fibers, as designed for high thermal conductivity, low thermal expansion, low dielectric constant, high/low electrical conductivity, and electromagnetic interference shielding. Applications include heat sinks, housings, printed wiring boards, substrates, lids, die attach, encapsulation, interconnections, and thermal interface materials.

RELEVANT APPENDICES: *A, B, C*

3.1 INTRODUCTION

Composite materials are usually designed for use as structural materials. With the rapid growth of the electronics industry, they are finding more and more electronic applications. Due to the vast difference in property requirements between structural and electronic composites, the design criteria are different. While structural composites emphasize high strength and high modulus, electronic composites emphasize high thermal conductivity, low thermal expansion, low dielectric constant, high/low

electrical conductivity, and/or electromagnetic interference (EMI) shielding effectiveness, depending on the particular electronic application. Low density is desirable
for both aerospace structures and aerospace electronics. Structural composites stress
processability into large parts, such as panels, whereas electronic composites emphasize processability into small parts, such as stand-alone films and coatings. Because
of the small size of the parts, material costs tend to be less of a concern for electronic
composites than for structural composites. For example, electronic composites can
use expensive fillers such as silver particles, which provide high electrical conductivity.

3.2 APPLICATIONS IN MICROELECTRONICS

The applications of polymer-matrix composites in microelectronics include interconnections, printed circuit boards, substrates, encapsulations, interlayer dielectrics,
die attach, electrical contacts, connectors, thermal interface materials, heat sinks,
lids, and housings. In general, the integrated circuit chips are attached to a substrate
or a printed circuit board on which the interconnection lines have been written on
each layer of the multilayer substrate or board. To increase the interconnection
density, another multilayer involving thinner layers of conductors and interlayer
dielectrics may be applied to the substrate before the chip is attached. By means of
soldered joints, wires connect between electrical contact pads on the chip and
electrical contact pads on the substrate or board. The chip may be encapsulated with
a dielectric for protection. It may also be covered by a thermally conducting (metal)
lid. The substrate or board is attached to a heat sink. A thermal interface material
may then be placed between the substrate or board and the heat sink to enhance the
quality of the thermal contact. The whole assembly may be placed in a thermally
conducting housing.

A printed circuit board is a sheet for the attachment of chips, whether mounted
on substrates, chip carriers, or otherwise, and for drawing interconnections. It is a
polymer-matrix composite that is electrically insulating and has four conductor lines
(interconnections) on one or both sides. Multilayer boards have lines on each inside
layer so that interconnections on different layers may be connected by short conductor columns called electrical vias. Printed circuit boards for mounting pin-
inserting-type packages need to have lead insertion holes punched through them.
Printed circuit boards for surface-mounting-type packages need no holes. Surface-
mounting-type packages, whether with leads, leaded chip carriers, without leads, or
leadless chip carriers (LLCCs), can be mounted on both sides of a circuit board,
whereas pin-inserting-type packages can only be mounted on one side. In surface
mounting technology (SMT), the surfaces of conductor patterns are connected
together electrically without employing holes. Solder is used to make electrical
connections between a surface-mounting-type package (leaded or leadless) and a
circuit board. A lead insertion hole for pin-inserting-type packages is a plated-
through hole, a hole on whose wall a metal is deposited to form a conducting
penetrating connection. After pin insertion, the space between the wall and the pin
is filled by solder to form a solder joint. Another type of plated-through hole is a
via hole, which connects different conductor layers without the insertion of a lead.

A substrate, also called a chip carrier, is a sheet on which one or more chips are attached and interconnections are drawn. In the case of a multilayer substrate, interconnections are also drawn on each layer inside the substrate such that interconnections in different layers are connected, if desired, by electrical vias. A substrate is usually an electrical insulator. Substrate materials include ceramics (Al_2O_3, AlN, mullite, or glass ceramics), polymers (polyimide), semiconductors (silicon), and metals (aluminum). The most common substrate material is Al_2O_3. As the sintering of Al_2O_3 requires temperatures greater than 1000°C, the metal interconnections need to be refractory, such as tungsten or molybdenum. The disadvantages of tungsten and molybdenum lie in higher electrical resistivity compared to copper. In order to make use of more conductive metals (e.g., Cu, Au, and Ag-Pd) as the interconnections, ceramics that sinter at temperatures below 1000°C can be used in place of Al_2O_3. The competition between ceramics and polymers for substrates is increasingly keen. Ceramics and polymers are both electrically insulating; ceramics are advantageous in that they have higher thermal conductivity than polymers; polymers are advantageous in that they have lower dielectric constant than ceramics. A high thermal conductivity is attractive for heat dissipation; a low dielectric constant is appealing for a smaller capacitive effect, hence a smaller signal delay. Metals are desirable for their very high thermal conductivity compared to ceramics and polymers.

An interconnection is a conductor line for signal transmission, power, or ground. It is usually in the form of a thick film of thickness > 1 μm. It can be on a chip, a substrate, or a printed circuit board. The thick film is made by either screen printing or plating. Thick-film conductor pastes containing silver particles and glass frit (binder that functions by the viscous flow of glass upon heating) are widely used to form thick-film conductor lines on substrates by screen printing and subsequent firing. These films suffer from the reduction of electrical conductivity by the presence of the glass, and by the porosity in the film after firing. The choice of a metal in a thick-film paste depends on the need for withstanding air oxidation in the heating encountered in subsequent processing, which can be the firing of the green thick-film together with the green ceramic substrate (a process known as cofiring). It is during cofiring that bonding and sintering take place. Copper is an excellent conductor, but it oxidizes readily when heated in air. The choice of metal also depends on the temperature encountered in subsequent processing. Refractory metals, such as tungsten and molybdenum, are suitable for interconnections heated to high temperatures (> 1000°C), for example during Al_2O_3 substrate processing.

A z-axis anisotropic electrical conductor film is a film that is electrically conducting only in the z-axis, i.e., in the direction perpendicular to the plane of the film. As one z-axis film can replace a whole array of solder joints, z-axis films are valuable for solder replacement, processing cost reduction, and repairability improvement in surface-mount technology.

An interlayer dielectric is a dielectric film separating the interconnection layers such that the two kinds of layers alternate and form a thin-film multilayer. The dielectric is a polymer, usually spun on or sprayed, or a ceramic applied by chemical vapor deposition (CVD). The most common multilayer involves polyimide as the dielectric and copper interconnections, plated, sputtered, or electron-beam deposited.

A die attach is a material for joining a die (a chip) to a substrate. It can be a metal alloy (a solder paste), a polymeric adhesive (a thermoset or a thermoplast), or a glass. Die attach materials are usually applied by screen printing. A solder is attractive in its high thermal conductivity, which enhances heat dissipation. However, its application requires the use of heat and a flux. The flux subsequently needs to be removed chemically. The defluxing process adds cost and is undesirable for the environment (the ozone layer) due to the chlorinated chemicals used. A polymer or glass has poor thermal conductity, but this can be alleviated by the use of thermally conductive filler such as silver particles. A thermoplast provides a reworkable joint, whereas a thermoset does not. Furthermore, a thermoplast is more ductile than a thermoset. Moreover, a solder suffers from its tendency to experience thermal fatigue due to the thermal expansion mismatch between the chip and the substrate, and the resulting work hardening and cracking of the solder. In addition, the footprint left by a solder tends to be larger than the footprint left by a polymer because of the ease with which the molten solder flows.

An encapsulation is an electrically insulating conformal coating on a chip for protection against moisture and mobile ions. An encapsulation can be a polymer (epoxy, polyimide, polyimide siloxane, silicone gel, Parylene, or benzocyclobutene), which can be filled with SiO_2, BN, AlN, or other electrically insulating ceramic particles for decreasing the thermal expansion and increasing thermal conductivity.[1-3] The decrease of thermal expansion is needed because a neat polymer typically has a much higher coefficient thermal expansion than a semiconductor chip. An encapsulation can also be a ceramic (e.g., SiO_2, Si_3Ni_4, or silicon oxynitride). In electronic packaging, encapsulation is a step performed after both die bonding and wire bonding and before the packaging, using a molding material. The molding material is typically a polymer, such as epoxy. However, it can also be a ceramic, such as Si_3N_4, cordierite (magnesium silicate), SiO_2, etc. A ceramic is advantageous in its low coefficient of thermal expansion and higher thermal conductivity, but it is much less convenient to apply than a polymer.

A lid is a cover for a chip that offers physical protection. The chip is typically mounted in a well in a ceramic substrate and the lid covers the well. A lid is preferably a metal because of the need to dissipate heat. It is joined to the ceramic substrate by soldering, using a solder preform (e.g., Au-Sn) shaped like a gasket. Due to the low coefficient of thermal expansion, Kovar® (54Fe-29Ni-17Co) is used for the lid. For the same reason, Kovar is often used for the can in which a substrate is mounted. Although Kovar has a low coefficient of thermal expansion (5.3×10^{-6} °C^{-1} at 20–200°C), it also has a low thermal conductivity of 17 W m^{-1} K^{-1}.

A heat sink is a thermal conductor that conducts and radiates heat away from the circuitry. It is typically bonded to a printed circuit board. The thermal resistance of the bond and that of the heat sink govern the effectiveness of heat dissipation. A heat sink that matches the coefficient of thermal expansion of the circuit board is desirable for resistance to thermal cycling.

Insufficiently fast dissipation of heat is the most critical problem that limits the reliability and performance of microelectronics.[4] The problem becomes more severe as electronics are miniaturized. The problem also accentuates as power (voltage and current) increases. Excessive heating from insufficient heat dissipation causes thermal stress in the electronic package, which may cause warpage of the semiconductor chip. The problem is also compounded by thermal fatigue, which results from cyclic heating and thermal expansion mismatches. Since there are many solder joints in an electronic package, solder joint failure is a big reliability concern. For these reasons, thermal management has become a key issue within the field of electronic packaging. Thermal management refers to the use of materials, devices, and packaging schemes to attain efficient heat dissipation.

The use of materials with high thermal conductivity and low thermal expansion for heat sinks, lids, housings, substrates, and die attach is an important avenue for alleviating the heat dissipation problem. For this purpose, metal-matrix composites (such as silicon carbide particle aluminum-matrix composites) and polymer-matrix composites (such as silver particle-filled epoxy) have been developed. However, another method, which has received much less attention, is the improvement of the thermal contact between the various components in an electronic package. The higher thermal conductivity of the individual components cannot effectively help heat dissipation unless thermal contacts between the components are good. Without good thermal contacts, the use of expensive thermal conducting materials is a waste.

The improvement of a thermal contact involves the use of a thermal interface material, such as a thermal fluid, a thermal grease, or a resilient thermal conductor. A thermal fluid or grease is spread on the mating surfaces. A resilient thermal conductor is sandwiched by the mating surfaces and held in place by pressure. A thermal fluid is most commonly mineral oil. Thermal greases are usually conducting particle-filled silicone. Resilient thermal conductors are conducting particle-filled elastomers. Of these three types of thermal interface materials, thermal greases are by far the most common. Resilient thermal conductors are not as well developed as thermal fluids or greases.

As the materials to be interfaced are good thermal conductors, the effectiveness of a thermal interface material is enhanced by high thermal conductivity and low thickness of the interface material, and low thermal contact resistance between the interface material and each mating surface. As the mating surfaces are not perfectly smooth, the interface material must be able to flow or deform so as to conform to their topology. If the interface material is a fluid, grease, or paste, it should have a high fluidity in order to conform and to have a small thickness after mating. On the other hand, the thermal conductivity of the grease or paste increases with increasing filler content, and this is accompanied by decrease in workability. Without a filler, as in the case of an oil, thermal conductivity is poor. A thermal interface material in the form of a resilient thermal conductor sheet cannot usually be as thin or conformable as one in the form of a fluid, grease, or paste, so its effectiveness requires a very high thermal conductivity.

3.3 POLYMER-MATRIX COMPOSITES

Polymer-matrix composites with continuous or discontinuous fillers are used for electronic packaging and thermal management. Composites with continuous fillers are used as substrates, heat sinks, and enclosures. Composites with discontinuous fillers are used for die attach, electrically/thermally conducting adhesives, encapsulations, thermal interface materials, and electrical interconnections. Composites with discontinuous fillers can be in a paste form during processing, thus allowing application by printing and injection molding. Composites with continuous fillers cannot undergo paste processing, but the continuous fillers provide lower thermal expansion and higher conductivity than discontinuous fillers.

Composites can have thermoplastic or thermosetting matrices. Thermoplastic matrices have the advantage that a connection can be reworked by heating for the purpose of repair, whereas thermosetting matrices do not allow reworking. On the other hand, controlled-order thermosets are attractive for their thermal stability and dielectric properties.[5] Polymers exhibiting low dielectric constant, low dissipation factor, low coefficient of thermal expansion, and compliance are preferred.[6]

Composites can be electrically conducting or electrically insulating; the electrical conductivity is provided by a conductive filler. The composites can be both electrically and thermally conducting, as attained by the use of metal or graphite fillers. They can also be electrically insulating but thermally conducting, as conveyed by the use of diamond, aluminum nitride, boron nitride, or alumina fillers.[7,8] An electrically conducting composite can be isotropically conducting[9,10] or anisotropically conducting.[11] A z-axis conductor is an example of an anisotropic conductor.

3.3.1 POLYMER-MATRIX COMPOSITES WITH CONTINUOUS FILLERS

Epoxy-matrix composites with continuous glass fibers and made by lamination are used for printed wiring boards because of the electrically insulating property of glass fibers, and the good adhesive behavior and established industrial usage of epoxy. Aramid (Kevlar™) fibers can be used instead of glass fibers to provide lower dielectric constant.[12] Alumina (Al_2O_3) fibers can be used for increasing the thermal conductivity.[13] By selecting the fiber orientation and loading in the composite, the dielectric constant can be decreased and the thermal conductivity increased.[14] By impregnating the yarns or fabrics with a silica-based sol and subsequent firing, the thermal expansion can be reduced.[15] Matrices other than epoxy can be used, such as polyimide and cyanate ester.[16]

For heat sinks and enclosures, conducting fibers are used since they enhance thermal conductivity and the ability to shield electromagnetic interference. EMI shielding is particularly important for enclosures.[17] Carbon fibers are most commonly used for these applications due to their conductivity, low thermal expansion, and wide availability as a structural reinforcement. For high thermal conductivity, carbon fibers made from mesophase pitch[18-24] or copper plated carbon fibers are preferred.[25-27] For EMI shielding, both uncoated carbon fibers[28,29] and metal-coated carbon fibers[30,31] have been used.

For avionic electronic enclosures, low density is essential for saving aircraft fuel. Aluminum is the traditional material for this application. Because of mechanical, electrical, environmental, manufacturing/producibility, and design-to-cost criteria, carbon fiber reinforced epoxy has been judged more attractive than aluminum, glass fiber reinforced epoxy, glass fiber reinforced epoxy with aluminum interlayer, beryllium, aluminum-beryllium, and SiC particle reinforced aluminum.[32] A related application is thermal management of satellites, for which the thermal management materials need to be integrated from the satellite structure down to the electronic device packaging.[33] Continuous carbon fibers are suitable for this application due to their high thermal conductivity, low density, high strength, and high modulus.

3.3.2 POLYMER-MATRIX COMPOSITES WITH DISCONTINUOUS FILLERS

Polymer-matrix composites with discontinuous fillers are widely used in electronics,[34] despite their poor mechanical properties compared to composites with continuous fibers. This is because materials in electronics do not need to be mechanically strong and discontinuous fillers enable processing through the paste form that is particularly suitable for making films, whether standalone or on a substrate.

Screen printing is a common method for patterning a film on a substrate. In the case of an electrically conducting paste, the pattern is an array of electrical interconnections and electrical contact pads on the substrate. As screen printing involves the paste going through a screen, screen printable pastes contain particles and no fiber, and the particles must be sufficiently small, typically less than 10 μm. The larger the particles, the poorer the patternability, i.e., the edge of a printed line is not sufficiently well defined. In applications not requiring patternability, such as thermal interface materials, short fibers are advantageous in that the connectivity of the short fibers is superior to that of particles at the same volume fraction. For a conducting composite, better connectivity of the filler units means higher conductivity for the composite. Instead of using short fibers, one may use elongated particles or flakes for the sake of the connectivity. In general, the higher the aspect ratio, the better the connectivity for the same volume fraction. The use of elongated particles or flakes can provide an aspect ratio larger than one, while retaining patternability. It is an attractive compromise.

In case of a conducting composite, the greater the volume fraction of the conducting filler, the higher the conductivity of the composite, since the polymer matrix is insulating. However, the greater the filler volume fraction, the higher the viscosity of the paste, and the poorer its processability. To attain a high filler volume fraction while maintaining processability, a polymer of low viscosity is preferred, and good wettability of the filler by the matrix, as provided by filler surface treatments and/or the use of surfactants, is desirable.

The matrix used in making a polymer-matrix composite can be in the form of a liquid (a thermosetting resin) or a solid (a thermoplastic powder) during the mixing of the matrix and the filler. In the case of the matrix in the form of a powder, the distribution of the filler units in the resulting composite depends on the size of the matrix powder particles; the filler units line the interface between adjacent matrix particles and the filler volume fraction needed for percolation (i.e., the filler units touching one another to form a continuous path)[35] decreases with increasing matrix particle size. The reaching of percolation is accompanied by a large increase in conductivity. However, a large matrix particle size is detrimental to processability, so a compromise is needed.

In the case of the matrix in the form of a thermoplastic powder, the percolation attained may be degraded or destroyed after subsequent composite fabrication involving flow of the thermoplastic under heat and pressure. In this case, a thermo-plastic that flows less is preferred for attaining high conductivity in the resulting composite.[36]

A less common way to attain percolation in a given direction is to apply an electric or magnetic field to align the filler units along the direction. For this technique to be possible, the filler units must be polarizable electrically or magnet-ically. Such alignment is one of the techniques used to produce z-axis conductors.

In percolation, the filler units touch one another to form continuous paths, but there is considerable contact resistance at the interfaces between them. To decrease this contact resistance, thereby increasing the conductivity of the composite, one can increase the size of the filler units so the amount of interface area is decreased, provided percolation is maintained. A less common but more effective way is to bond the filler units together at their junctions by using a solid (like solder) that melts and wets the surface of the filler during composite fabrication. The low-melting-point solid can be in the form of particles added to the composite mix, or in the form of a coating on the filler units. In this way, a three-dimensionally interconnected conducting network is formed after composite fabrication.[37]

An intimate interface between the filler and the matrix is important to the conductivity of a composite, even though the filler is conducting and the matrix may be perfectly insulating. This is because conduction may involve a path from a filler unit to an adjacent one through a thin film of the matrix by means of tunneling. In the case of the matrix being slightly conducting (but not as conducting as the filler), the conduction path involves both the filler and the matrix, and the filler-matrix interface is even more important. This interface may be improved by filler surface treatments prior to incorporating the filler in the composite, or by the use of a surfactant.[38]

The difference in thermal expansion coefficient between filler and matrix, and the fact that composite fabrication occurs at an elevated temperature, cause thermal stress during cooling of the fabricated composite. The thermal expansion coefficient of a polymer is relatively high, so the filler units are usually under compression after cooling. Compression helps to tighten the filler-matrix interface, though the com-pressive stress in the filler and the tensile stress in the matrix may degrade the performance and durability of the composite.

In case of the matrix being conducting, but not as conducting as the filler, as for conducting polymer matrices,[39,40] percolation is not essential for the composite to be conducting, though percolation would greatly enhance conductivity. Below the percolation threshold (the filler volume fraction above which percolation occurs), the conductivity of the composite is enhanced by a uniform distribution of the filler units since the chance increases of having a conduction path that involves more filler and less matrix as filler distribution becomes more uniform. Uniformity is never perfect; it is described by the degree of dispersion of the filler. The degree of dispersion can be enhanced by rigorous agitation during mixing of the filler and matrix, or by the use of a dispersant. In the case of a matrix in the form of particles that are coarser than the filler units, the addition of fine particles to the mix also helps dispersion of the filler.[41]

Because the thermal expansion coefficient of a polymer is relatively high, the polymer matrix expands more than the filler during the heating of a polymer-matrix composite. This results in the proximity between adjacent filler units changing with temperature, decreasing the conductivity of the composite.[42] This phenomenon is detrimental to the thermal stability of composites.

Corrosion and surface oxidation of the filler are the most common causes of degradation that decreases the conductivity of the composite. Thus, oxidation-resistant fillers are essential. Silver and gold are oxidation resistant, but copper is not. Due to the high cost of silver and gold, the coating of copper, nickel, or other lower-cost metal fillers by gold or silver is common for improving oxidation resistance. The most common filler by far is silver particles.[43,44]

A z-axis anisotropic electrical conductor film is electrically conducting in the direction perpendicular to the film, but is insulating in all other directions. This film is technologically valuable for use as an interconnection material in electronic packaging, as it electrically connects the electrical contact pads touching one side of the film with the corresponding contact pads touching the direct opposite side of the film. Even though the film is in one piece, it contains numerous z-axis conducting paths, so it can provide numerous interconnections. If each contact pad is large enough to span a few z-axis conducting paths, no alignment is needed between the contact pad array and the z-axis film, whether the conducting paths are ordered or random in their distribution.[45-59] In this situation, in order to attain a high density of interconnections, the cross-section of each z-axis conducting path must be small. However, if each contact pad is only large enough to span one z-axis conducting path, alignment is needed between the contact pad array and the z-axis film, and this means the conducting paths in the z-axis film must be ordered in the same way as the contact pad array.[60]

An example of an application of a z-axis conductor film is in the interconnections between the leads from (or contact pads on) a surface-mount electronic device and the contact pads on the substrate beneath the device. In this application, one piece of z-axis film can replace a whole array of solder joints, so the processing cost can be reduced. Furthermore, the problem of thermal fatigue of the solder joints can be avoided by this replacement. Another example is in the vertical interconnections in three-dimensional electronic packaging.

A z-axis film is a polymer-matrix composite containing conducting units that form the z-axis conducting paths. The conducting units are usually particles, such as metal particles and metal-coated polymer particles. The particles can be clustered so that each cluster corresponds to one conducting path.[45-47,51] Metal columns,[52] metal particle columns (e.g., gold-plated nickel),[50,51] and individual metal-coated polymer particles[49] have been used to provide z-axis conducting paths. Particle columns were formed by magnetic alignment. Using particle columns, Reference 46 attained a conducting path width of 400 μm and a pitch (center-to-center distance between adjacent conducting paths) of 290 μm. Also using particle columns, References 50 and 51 attained a conducting path width of ~ 10 μm and a pitch of ~ 100 μm. In general, a large conducting path width is desirable for decreasing the resistance per path, while a small pitch is desirable for high density interconnection.

In contrast to the use of metal wires, metal columns, or metal particle columns, Reference 61 used one metal particle per conducting path. The concept of one particle per path had been demonstrated by Reference 49 by using metal-coated polymer particles. However, due to the high resistivity of the metal coating compared to the bulk metal, the z-axis resistivity of the film was high (0.5 Ω.cm for a conducting path). By using metal particle in place of metal coated polymer particles, Reference 61 decreased the z-axis resistivity of a conducting path to 10^{-6} Ω.cm. Furthermore, Reference 61 did not rely on a polymer for providing resilience; resilience is provided by the metal particles, which protrude from both sides of the stand-alone film. As a result, the problem of stress relaxation of the polymer is eliminated. In addition, the protrusion of the metal particles eliminates the problem of open-circuiting the connection upon heating because of the higher thermal expansion of the polymer compared to the conductor.[58]

Most work on z-axis adhesive films[56,57] used an adhesive with randomly dispersed conductive particles (8–12 μm diameter) suspended in it. The particles were phenolic spheres that had been coated with nickel. After bonding under heat (180–190°C) and pressure (1.9 MPa), a particle became oval in shape (4 μm thick). There was one particle per conducting path. The main drawback of this technology is the requirement of heat and pressure for curing the adhesive. Heat and pressure are not desirable in practical use of the z-axis adhesive. Reference 61 removed the need for heat and pressure through the choice of the polymer.

A different kind of z-axis adhesive film[60] used screening or stenciling to obtain a regular two-dimensional array of silver-filled epoxy conductive dots, but this technology suffers from the large pitch (1500 μm) of the dots and the consequent need for alignment between z-axis film and contact pad array. In the work of Reference 61, the pitch of the conducting paths in the z-axis adhesive film is as low as 64 μm.

Capacitors require materials with a high dielectric constant. Such materials in the form of thick films allow capacitors to be integrated with the electronic packaging, allowing further miniaturization and performance and reliability improvements.[62] These thick-film pastes involve ceramic particles with a high dielectric constant, such as barium titanate ($BaTiO_3$), and a polymer (e.g., epoxy).[63,64]

Inductors are needed for transformers, DC/DC converters, and other power supply applications. They require magnetic materials. Such materials in the form of thick films allow inductors and transformers to be integrated with the electronic packaging, allowing further miniaturization. These thick-film pastes involve magnetic particles (e.g., ferrite) and a polymer.[65,66]

The need for EMI shielding is increasing rapidly due to the interference of radio frequency radiation (such as that from cellular phones) with digital electronics, and society's increasing dependence on digital electronics. The associated electronic pollution is an interference problem.

EMI shielding is achieved by using electrical conductors, such as metals and conductive-filled polymers.[67-75] EMI-shielding gaskets[76-87] are resilient conductors. They are needed to electromagnetically seal an enclosure. The resilient conductors are most commonly elastomers (rubber) that are filled with a conductive filler,[88] or elastomers that are coated with a metalized layer. Metallized elastomers suffer from poor durability because of the tendency of the metal layer to debond from the elastomer. Conductive-filled elastomers do not have this problem, but they require the use of a highly conductive filler, such as silver particles, in order to attain a high shielding effectiveness while maintaining resilience. The highly conductive filler is expensive, which makes the composite expensive. The use of a less conducting filler results in the need for a large volume fraction of the filler in order to attain a high shielding effectiveness; the consequence is diminished resilience or even loss of resilience. Moreover, these composites suffer from degradation of shielding effectiveness in the presence of moisture or solvents. In addition, the polymer matrix in the composites limits the temperature resistance, and the thermal expansion mismatch between filler and matrix limits the thermal cycling resistance.

Due to the skin effect (electromagnetic radiation at a high frequency interacting with only the near surface region of an electrical conductor), a filler for a polymer-matrix composite for EMI shielding needs to be small in unit size. Although connectivity between the filler units is not required for shielding, it helps. Therefore, a filler in the form of a metal fiber of very small diameter is desirable. For this purpose, nickel filaments of diameter 0.4 μm and length > 100 μm, with a carbon core of diameter 0.1 μm, were developed.[89] Their exceptionally small diameters compared to those of existing metal fibers made them outstanding for use as fillers in polymers for EMI shielding. A shielding effectiveness of 87 dB at 1 GHz was attained in a polyethersulfone-matrix composite with only 7 vol.% nickel filaments.[89-91] The low volume fraction allows resilience in a silicone-matrix composite for EMI gaskets.[92]

3.4 SUMMARY

Polymer-matrix composite materials for microelectronics are designed for high thermal conductivity, low CTE, low dielectric constant, either high or low electrical conductivity, and processability (e.g., printability). Applications include heat sinks, housings, printed wiring boards, substrates, lids, die attach, encapsulation, interconnections, thermal interface materials, and EMI shielding. Combinations of properties are usually required. For example, for heat sinks and substrates, the combination of high thermal conductivity and low CTE is required for heat dissipation and thermal stress reduction. In the case of aerospace electronics, low density is also desired. Polymer-matrix composites for microelectronics include those with continuous and discontinuous fillers. They can be in the form of an adhesive film, a stand-alone film, or a bulk material, and can be electrically isotropic or anisotropic.

REFERENCES

1. V. Sarihan and T. Fang, Structural Analysis in Microelectronic and Fiber Optic Systems, *EEP*, Vol. 12, pp. 1-4 (1995).
2. W.E. Marsh, K. Kanakarajan and G.D. Osborn, Polymer/Inorganic Interfaces, Materials Research Soc. Symp. Proc., Materials Research Soc., Vol. 304, pp. 91-96 (1993).
3. L.T. Nguyen and I.C. Noyan, *Polymer Eng. Sci.*, 28(16), 1013-1025 (1988).
4. W.-J. Yang and K. Kudo, Proc. Int. Symp. Heat Transfer Science and Technology, pp. 14-30 (1988).

5. H. Korner, A. Shiota and C.K. Ober, ANTEC '96: Plastics — Racing into the Future, *Conf. Proc., Society of Plastics Engineers*, Technical Papers Series, no. 42, vol. 2, pp. 1458-1461 (1996).

6. G.S. Swei and D.J. Arthur, 3rd Int. SAMPE Symp. Exhib., pp. 1111-1124 (1989).

7. L. Li and D.D.L. Chung, *Composites*, 22(3),211-218 (1991).

8. X. Lu and G. XU, *J. Appl. Polymer Sci.*, 65(13), 2733-2738 (1997).

9. D. Klosterman and L. LI, *J. Electron. Manuf.*, 5(4), 277-287 (1995).

10. S.K. Kang, R. Rai and S. Purushothaman, 1996 Proc. 46th Electronic Components & Technology, IEEE, pp. 565-570 (1996).

11. G.F.C.M. Lijten, H.M. Van Noort and P.J.M. Beris, *J. Electron. Manuf.*, 5(4), 253-261 (1995).

12. M.P. Zussman, B. Kirayoglu, S. Sharkey and D.J. Powell, 6th Int. SAMPE Electronics Conf., pp. 437-448 (1992).

13. J.D. Bolt and R.H. French, *Adv. Mater. Process.*, 134(1), 32-35 (1988).

14. J.D. Bolt, D.P. Button and B.A. Yost, *Mater. Sci. Eng.*, A109, 207-211 (1989).

15. S.P. Mukerherjee, D. Suryanarayana and D.H. Strope, *J. Non-Crystalline Solids*, 147-148, 783-791 (1992).

16. M.P. Zussman, B. Kirayoglu, S. Sharkey and D.J. Powell, 6th Int. SAMPE Electronics Conf., pp. 437-448 (1992).

17. J.J. Glatz, R. Morgan and D. Neiswinger, 6th Int. SAMPE Electronics Conf., Vol. 6, pp. 131-145 (1992).

18. A. Bertram, K. Beasley and W. De La Torre, *Nav. Eng. J.*, 104(3), 276-285 (1992).

19. D. Brookstein and D. Maass, 7th Int. SAMPE Electronics Conf., pp. 310-327 (1994).

20. T.F. Fleming and W.C. Riley, *Proc. SPIE*, 1997, 136-147 (1993).

21. T.F. Fleming, C.D. Levan and W.C. Riley, Proc. Technical Conf., Int. Electronics Packaging Conf., pp. 493-503 (1995).

22. A.M. Ibrahim, 6th Int. SAMPE Electronics Conf., pp. 556-567 (1992).

23. N. Kiuchi, K. Ozawa, T. Komami, O. Katoh, Y. Arai, T. Watanabe and S. Iwai, Int. SAMPE Tech. Conf., Vol. 30, pp. 68-77 (1998).

24. J.W.M. Spicer, D.W. Wilson, R. Osiander, J. Thomas and B.O. Oni, *Proc. SPIE*, 3700, 40-47 (1999).

25. D.A. Foster, *SAMPE Q.*, 21(1), 58-64 (1989).

26. D.A. Foster, 34th Int. SAMPE Symp. Exhib., Book 2, pp. 1401-1410 (1989).

27. W. De La Torre, 6th Int. SAMPE Electronics Conf., pp. 720-733 (1992).

28. P.D. Wienhold, D.S. Mehoke, J.C. Roberts, G.R. Seylar and D.L. Kirkbride, Int. SAMPE Tech. Conf., pp. 243-255 (1998).

29. X. Luo and D.D.L. Chung, *Composites*: Part B, 30(3), 227-231 (1999).

30. L.G. Morin Jr. and R.E. Duvall, Proc. Int. SAMPE Symp. Exhib., 43(1), 874-881 (1998).

31. G. Lu, X. Li and H. Jiang, *Composites Sci. Tech.*, 56, 193-200 (1996).

32. P.L. Smaldone, 27th Int. SAMPE Technical Conf., pp. 819-829 (1995).

33. J.J. Glatz, D.L. Vrable, T. Schmedake and C. Johnson, 6th Int. SAMPE Electronics Conf., pp. 334-346 (1992).

34. R. Crossman, Northcon/85 — Conf. Rec. (1985).

35. D.S. McLachlan, M. Blaszkiewicz and R.E. Newnham, *J. Am. Ceram. Soc.*, 73(8), 2187-2203 (1990).

36. L. Li and D.D.L. Chung, *Polym. Compos.*, 14(6), 467-472 (1993).

37. L. Li, P. Yih and D.D.L. Chung, *J. Electron. Mater.*, 21(11), 1065-1071 (1992).

38. B. Guerrero, C. Alemán and R. Garza, *J. Polym. Eng.*, 17(2), 95-110 (1997-1998).

39. M. Omastová, J. Pavlinec, J. Pionteck and F. Simon, *Polym. Int.*, 43(2), 109-116 (1997).

40. X.B. Chen and J.-P. Issi, M. Cassart, J. Devaux and D. Billaud, *Polymer,* 35(24), 5256-5258 (1994).
41. P. Chen and D.D.L. Chung, *J. Electron. Mater.,* 24(1), 47-51 (1995).
42. J. Fournier, G. Boiteux, G. Seytre and G. Marichy, *J. Mater. Sci. Lett.,* 16(20), 1677-1679 (1997).
43. D. Klosterman and L. Li, *J. Electron. Manuf.,* 5(4), 277-287 (1995).
44. S.K. Kang, R. Rai and S. Purushothaman, Proc. 46th Electronic Components & Technology, IEEE, pp. 565-570 (1996).
45. W.R. Lambert and W.H. Knausenberger, Proceedings of the Technical Program — National Electronic Packaging and Production Conference, 3, 1512-1526 (1991).
46. J.A. Fulton, D.R. Horton, R.C. Moore, W.R. Lambert and J.J. Mottine, Proc. 39th Electronic Components Conf., IEEE, pp. 71-77 (1989).
47. W.R. Lambert, J.P. Mitchell, J.A. Suchin and J.A. Fulton, Proc. 39th Electronic Components Conf., IEEE, pp. 99-106 (1989).
48. P.B. Hogerton, J.B. Hall, J.M. Pujol and R.S. Reylek, *Mat. Res. Soc. Symp. Proc.,* 154, 415 (1989).
49. L. Li and D.D.L. Chung, *J. Electron. Packag.,* 119(4), 255 (1997).
50. P.T. Robinson, V. Florescu, G. Rosen and M.T. Singer, Annual Connector & Interconnection Technology Symposium, pp. 507-515 (1990).
51. G. Rosen, P.T. Robinson, V. Florescu and M.T. Singer, Proc. 6th IEEE Holm Conf. Electrical Contacts and 15th Int. Conf. Electric Contacts, IEEE, pp. 151-165 (1990).
52. D.D. Johnson, Proc. 3rd Int. Symp. Advanced Packaging Materials: Processes, Properties and Interfaces, IEEE, pp. 29-30 (1997).
53. T. Kokogawa, H. Morishita, K. Adachi, H. Otsuki, H. Takasago and T. Yamazaki, Conference Record of 1991 International Display Confrence, IEEE, and Society for Information Display, pp. 45-48 (1991).
54. N.P. Kreutter, B.K. Grove, P.B. Hogerton and C.R. Jensen, 7th Electronic Materials and Processing Congress (Aug. 1992).
55. H. Yoshigahara, Y. Sagami, T. Yamazaki, A. Burkhart and M. Edwards, Proc. Technical Program: National Electronic Packaging and Production Conf., NEPCON West '91, 1, 213-219 (1991).
56. D.M. Bruner, *Int. J. Microcircuite Electron. Packag.,* 18(3), 311 (1995).
57. J.J. CREA and P.B. Hogerton, Proc. Technical Program: National Electronic Packaging and Production Conf., NEPCON West '91, 1, 251-259 (1991).
58. K. Gilleo, Proc. Electricon '94, Electronics Manufacturing Productivity Facility, pp. 11/1-11/12 (1994).
59. K. Chung, G. Dreier, P. Fitzgerald, A. Boyle, M. Lin and J. Sager, Proc. 41st Electronic Components Conf., p. 345 (May 1991).
60. J.C. Bolger and J.M. Czarnowski, 1995 Japan IEMT Symp. Proc., 1995 Int. Electronic Manufacturing Technology Symp., IEEE, pp. 476-481 (1996).
61. Y. Xu and D.D.L. Chung, *J. Electron. Mater.,* 28(11), 1307 (1999).
62. G.Y. Chin, *Adv. Mater. Proc.,* 137(1), 47, 50, 86 (1990).
63. S. Liang, S.R. Chong and E.P. Giannelis, Proc. Electronic Components & Technology Conf., IEEE, Vol. 48, pp. 171-175 (1998).
64. V. Agarwal, P. Chahal, R. R. Tummala and M.G. Allen, Proc. Electronic Components and Technology Conf., IEEE, Vol. 48, pp. 165-170 (1998).
65. J.Y. Park and M.G. Allen, Conf. Proc. — IEEE Applied Power Electronics Conf. and Exposition — APEC, Vol. 1, pp. 361-367 (1997).
66. J.Y. Park, L.K. Lagorce and M.G. Allen, *IEEE Trans. Magnetics,* 33(5), 3322-3324 (1997).

67. R. Charbonneau, Proc. Int. SAMPE Symp. Exhib., Vol. 43, No. 1, pp. 833-844 (1998).

68. J. Wang, V.V. Varadan and V.K. Varadan, *SAMPE J.,* 32(6), 18-22 (1996).

69. S. Maugdal and S. Sankaran, Recent Trends in Carbon, Proc. National Conf., O.P. Bahl, Ed., pp. 12-19 (1997).

70. J.T. Hoback and J.T. Reilly, *J. Elastomers Plast.,* 20(1), 54-69 (1998).

71. G. Lu, X. Li and H. Jiang, *Compos. Sci. Tech.,* 56, 193-200 (1996).

72. D.W. Radford, *J. Adv. Mater.,* pp. 45-53 (Oct. 1994).

73. C. Huang and J. Pai, *J. Appl. Polym. Sci.,* 63(1), 115-123 (1997).

74. D.W. Radford and B.C. Cheng, *J. Test. Evaluation,* 21(5), 396-401 (1993).

75. S.R. Gerteisen, Northcon/85 — Conf. Rec., Paper 6.1 (1985).

76. P. O'Shea, *Evaluation Eng.,* 34(8), 84-93 (1995).

77. P. O'Shea, *Evaluation Eng.,* 35(8), 56-61 (1996).

78. R.A. Rothenberg and D.C. Inman, 1994 Int. Symp. Electromagnetic Compatibility, p. 818 (1994).

79. J.W.M. Child, *Electron. Prod.,* pp. 41-47 (Oct. 1986).

80. A.K. Subramanian, D.C. Pande and K. Boaz, Proc. 1995 Int. Conf. Electromagnetic Interference and Compatibility, Soc. EMC Engineers, pp. 139-147 (1995).

81. W. Hoge, *Evaluation Eng.,* 34(1), 84-86 (1995).

82. J.F. Walther, IEEE 1989 Int. Symp. Electromagnetic Compatibility: Symp. Record, pp. 40-45 (1989).

83. H.W. Denny and K.R. Shouse, IEEE 1990 Int. Symp. Electromagnetic Compatibility: Symp. Record, pp. 20-24 (1990).

84. R. Bates, S. Spence, J. Rowan and J. Hanrahan, 8th Int. Conf. Electromagnetic Compatibility, Electronics Division, Insitution of Electrical Engineers, pp. 246-250 (1992).

85. J.A. Catrysse, 8th Int. Conf. Electromagnetic Compatibility, Electronics Division, Institution of Electrical Engineers, pp. 251-255 (1992).

86. A.N. Faught, IEEE Int. Symp. Electromagnetic Compatibility, pp. 38-44 (1982).

87. G. Kunkel, IEEE Int. Symp. Electromagnetic Compatibility, pp. 211-216 (1980).

88. K.P. Sau, T.K. Chaki, A. Chakraborty and D. Khastgir, *Plast., Rubber Compos. Process Appl.,* 26(7), 291-297 (1997).

89. X. Shui and D.D.L. Chung, *J. Electron. Mater.,* 24(2), 107-113 (1995).

90. X. Shui and D.D.L. Chung, *J. Electron. Mater.,* 25(6), 930-934 (1996).

91. X. Shui and D.D.L. Chung, *J. Electron. Mater.,* 26(8), 928-934 (1997).

92. X. Shui and D.D.L. Chung, *J. Electron. Packag.,* 119(4), 236-238 (1997).

4 Materials for Electromagnetic Interference Shielding

CONTENTS

SYNOPSIS   Materials for the electromagnetic interference (EMI) shielding of electronics and radiation sources are reviewed, with emphasis on composite materials and resilient EMI gasket materials which shield mainly by reflection of the radiation at a high frequency.

RELEVANT APPENDICES: *E, D, C*

4.1   INTRODUCTION

EMI shielding refers to the reflection and/or adsorption of electromagnetic radiation by a material that acts as a shield against its penetration. As electromagnetic radiation, particularly at high frequencies (radio waves, such as those emanating from cellular phones) tend to interfere with electronics (computers), EMI shielding of both electronics and radiation sources is needed, and is increasingly required by governments around the world. The importance of EMI shielding relates to high demand on the reliability of electronics, and the rapid growth of radio frequency radiation sources.[1-9]

EMI shielding is distinguished from magnetic shielding, which refers to the shielding of magnetic fields at low frequencies (60 Hz). Materials for EMI shielding are different from those for magnetic shielding.

4.2 MECHANISMS OF SHIELDING

The primary mechanism of EMI shielding is reflection. For reflection of radiation, the shield must have mobile charge carriers (electrons or holes) that interact with the electromagnetic fields in the radiation. As a result, the shield tends to be electrically conducting, although a high conductivity is not required. For example, a volume resistivity of the order of 1 Ω.cm is usually sufficient. However, electrical conductivity is not the scientific criterion for shielding, as conduction requires connectivity in the conduction path, whereas shielding does not. Although shielding does not require connectivity, it is enhanced by it. Metals are by far the most common materials for EMI shielding. They function by reflection due to the free electrons in them. Metal sheets are bulky, so metal coatings made by electroplating, electroless plating, or vacuum deposition are used for shielding.[10-25] The coating may be on bulk materials, fibers, or particles, but coatings tend to suffer from their poor wear or scratch resistance.

A secondary mechanism of EMI shielding is absorption. For significant absorption of radiation, the shield should have electric and/or magnetic dipoles that interact with the electromagnetic fields in radiation. The electric dipoles may be provided by $BaTiO_3$ or other materials having a high value of the dielectric constant. The magnetic dipoles may be provided by Fe_3O_4 or other materials with a high value of magnetic permeability,[10] which may be enhanced by reducing the number of magnetic domain walls through the use of a multilayer of magnetic films.[26,27]

The absorption loss is a function of the product $\sigma_r\mu_r$, whereas the reflection loss is a function of the ratio σ_r/μ_r, where σ_r is the electrical conductivity relative to copper and μ_r is the relative magnetic permeability. Table 4.1 provides these factors for various materials. Silver, copper, gold, and aluminum are excellent for reflection because of their high conductivity. Superpermalloy and mumetal are excellent for absorption because of their high magnetic permeability. The reflection loss decreases with increasing frequency, whereas the absorption loss increases with increasing frequency.

Other than reflection and absorption, a mechanism of shielding is multiple reflections, the reflections at various surfaces or interfaces in the shield. This mechanism requires the presence of a large surface area or interface area. An example of a shield with a large surface area is a porous or foam material. An example of a shield with a large interface area is a composite material containing a filler that has a large surface area. The loss due to multiple reflections can be neglected when the distance between the reflecting surfaces or interfaces is large compared to the skin depth.

The losses, whether due to reflection, absorption, or multiple reflections, are expressed in dB. The sum of all the losses is the shielding effectiveness (in dB). The absorption loss is proportional to the thickness of the shield.

TABLE 4.1
Electrical Conductivity Relative to Copper (σ_r) and Relative Magnetic Permeability (μ_r) of Selected Materials (from Ref. 119)

Material	σ_r	μ_r	$\sigma_r\mu_r$	σ_r/μ_r
Silver	1.05	1	1.05	1.05
Copper	1	1	1	1
Gold	0.7	1	0.7	0.7
Aluminum	0.61	1	0.61	0.61
Brass	0.26	1	0.26	0.26
Bronze	0.18	1	0.18	0.18
Tin	0.15	1	0.15	0.15
Lead	0.08	1	0.08	0.08
Nickel	0.2	100	20	2×10^{-3}
Stainless steel (430)	0.02	500	10	4×10^{-5}
Mumetal (at 1 kHz)	0.03	20,000	600	1.5×10^{-6}
Superpermalloy (at 1 kHz)	0.03	100,000	3,000	3×10^{-7}

Electromagnetic radiation at high frequencies penetrates only the near surface region of an electrical conductor. This is known as the "skin effect." The electric field of a plane wave penetrating a conductor drops exponentially with increasing depth into the conductor. The depth at which the field drops to 1/e of the incident value is the skin depth (δ), which is given by

$$\delta = \frac{1}{\sqrt{\pi f \mu \sigma}} \tag{4.1}$$

where

f	= frequency,
μ	= magnetic permeability = $\mu_0\mu_r$,
μ_r	= relative magnetic permeability,
μ_0	= $4\pi \times 10^{-7}$ H/m, and
σ	= electrical conductivity in $\Omega^{-1}\text{m}^{-1}$.

Hence, the skin depth decreases with increasing frequency, and with increasing conductivity or permeability. For copper, $\mu_r = 1$, $\sigma = 5.8 \times 10^7$ $\Omega^{-1}\text{m}^{-1}$, so δ is 2.09 μm at a frequency of 1 GHz. For nickel of $\mu_r = 100$, $\sigma = 1.15 \times 10^7$ $\Omega^{-1}\text{m}^{-1}$, so δ is 0.47 μm at 1 GHz. The small value of δ for nickel compared to copper is due to the ferromagnetic nature of nickel.

4.3 COMPOSITE MATERIALS FOR SHIELDING

Due to the skin effect, a composite material having a conductive filler with a small unit size is more effective than one having a conductive filler with a large unit size. For effective use of the entire cross-section of a filler unit for shielding, the unit size of the filler should be comparable to or less than the skin depth. A filler of unit size 1 μm or less is typically preferred, though such a small unit size is not commonly available for most fillers, and the dispersion of the filler is more difficult when the filler unit size decreases. Metal-coated polymer fibers or particles are used as fillers for shielding, but they suffer from the fact that the polymer interior of each fiber or particle does not contribute to shielding.

Polymer-matrix composites containing conductive fillers are attractive for shielding[28-59] due to their processability (moldability), which helps to reduce or eliminate the seams in the housing that is the shield. The seams are commonly encountered in the case of metal sheets as the shield, and they tend to cause leakage of radiation and diminish the effectiveness of the shield. In addition, polymer-matrix composites are attractive in their low density. The polymer matrix is electrically insulating and does not contribute to shielding, though the polymer matrix can affect the connectivity of the conductive filler, and connectivity enhances shielding effectiveness. In addition, the polymer matrix affects processability.

Electrically conducting polymers[60-79] are becoming increasingly available, but they are not common and tend to be poor in processability and mechanical properties. Nevertheless, electrically conducting polymers do not require a conductive filler to provide shielding, so that they may be used with or without one. In the presence of a conductive filler, an electrically conducting polymer matrix has the added advantage of being able to electrically connect the filler units that do not touch one another, thereby enhancing connectivity.

Cement is slightly conducting, so the use of a cement matrix allows the conductive filler units in the composite to be electrically connected even when the filler units do not touch one another. Thus, cement-matrix composites have higher shielding effectiveness than corresponding polymer-matrix composites in which the polymer matrix is insulating.[80] Moreover, cement is less expensive than polymers, and cement-matrix composites are useful for shielding rooms in a building.[81-83] Similarly, carbon is a better matrix than polymers for shielding because of its conductivity, but carbon-matrix composites are expensive.[84]

A seam in a housing that serves as an EMI shield needs to be filled with an EMI gasket, which is commonly a material based on an elastomer, such as rubber.[85-98] An elastomer is resilient, but is not able to shield unless it is coated with a conductor (a metal coating called metallization) or is filled with a conductive filler (typically metal particles). The coating suffers from its poor wear resistance. The use of a conductive filler is problematic because of the resulting decrease in resilience, especially at the high filler volume fraction that is required for shielding effectiveness. Because the decrease in resilience becomes more severe as the filler concentration increases, a filler that is effective even at a low volume fraction is desirable. Therefore, the development of EMI gaskets is more challenging than that of EMI-shielding materials in general.

For a general EMI-shielding material in the form of a composite material, a filler that is effective at a low concentration is desirable, although that is not as critical as for EMI gaskets. This is because the strength and ductility of a composite decrease with increasing filler content when filler-matrix bonding is poor. Poor bonding is quite common for thermoplastic polymer matrices. Furthermore, a low filler content is desirable because of greater processability, which decreases with increasing viscosity. In addition, low filler content is desirable due to cost and weight saving.

In order for a conductive filler to be highly effective, it should have a small unit size (due to the skin effect), a high conductivity (for shielding by reflection and absorption), and a high aspect ratio (for connectivity). Metals are more attractive for shielding than carbons because of their higher conductivity, though carbons are attractive in their oxidation resistance and thermal stability. Fibers are more attractive than particles due to their high aspect ratio. Thus, metal fibers of a small diameter are desirable. Nickel filaments of diameter 0.4 μm have been shown to be particularly effective.[98-100] Nickel is more attractive than copper because of its superior oxidation resistance. Oxide film is poor in conductivity and is thus detrimental to connectivity among filler units.

Continuous fiber polymer-matrix structural composites that are capable of EMI shielding are needed for aircrafts and electronic enclosures.[84,101-109] The fibers in these composites are typically carbon fibers, which may be coated with a metal (nickel[110]) or be intercalated to increase conductivity.[111,112] An alternate design involves the use of glass fibers (not conducting) and conducting interlayers in the composite.[113,114]

4.4 EMERGING MATERIALS FOR SHIELDING

A particularly attractive EMI gasket material is flexible graphite, a flexible sheet made by compressing exfoliated graphite flakes (called worms) without a binder. During exfoliation, an intercalated graphite flake expands over 100 times along the c-axis. Compression of the resulting worms (like accordions) causes them to be mechanically interlocked so that a sheet is formed without a binder.

Due to exfoliation, flexible graphite has a large specific surface area (15 m²/g). Because of the absence of a binder, flexible graphite is essentially entirely graphite (other than the residual amount of intercalate in the exfoliated graphite). As a result, flexible graphite is chemically and thermally resistant, and low in coefficient of thermal expansion (CTE). Since its microstructure involves graphite layers that are parallel to the surface of the sheet, flexible graphite is high in electrical and thermal conductivities in the plane of the sheet. Because the graphite layers are somewhat connected perpendicular to the sheet (i.e., the honeycomb microstructure of exfoliated graphite), flexible graphite is electrically and thermally conductive in the direction perpendicular to the sheet (although not as conductive as the plane of the sheet). These in-plane and out-of-plane microstructures result in resilience, which is important for EMI gaskets. Due to the skin effect, a high surface is desirable for shielding. As the electrical conductivity and specific surface area are both quite high in flexible graphite, the effectiveness of this material for shielding is exceptionally high (up to 130 dB at 1 GHz).[115]

Other emerging materials for EMI shielding include woodceramics (porous carbons made by impregnating woody materials with phenol resin)[116,117] and aluminum honeycomb.[118]

4.5 CONCLUSION

Materials for EMI shielding are electrically conducting materials in the form of metals in bulk, porous and coating forms, composite materials with polymer, cement and carbon matrices, and carbons. In particular, EMI gasket materials, which require resilience, include elastomers and flexible graphite.

ACKNOWLEDGMENT

This work was supported by the U.S. Defense Advanced Research Projects Agency.

REFERENCES

1. D. Bjorklof, *Compliance Eng.,* 15(5), (1998).
2. R. Brewer and G. Fenical, *Evaluation Eng.,* 37(7), S-4–S-10 (1998).
3. P. O'Shea, *Evaluation Eng.,* 37(6), 40, 43, 45-46 (1998).
4. R.S.R. Devender, Proc. Int. Conf. Electromagnetic Interference and Compatibility, pp. 459-466 (1997).
5. B. Geddes, *Control,* 9(10), (1996).
6. S. Hempelmann, *Galvanotechnik,* 88(2), 418-424 (1997).
7. W.D. Kimmel and D.D. Gerke, *Medical Device & Diagnostic Industry,* 17(7), 112-115 (1995).
8. H.W. Markstein, *Electron. Packag. Prod.,* 35(2), (1995).
9. K.A. McRae, National Conf. Publication — Institution of Engineers, Australian, Vol. 2, No. 94/11, pp. 495-498 (1994).
10. V.V. Sadchikov and Z.G. Prudnikova, *Stal',* No. 4, pp. 66-69 (1997).
11. S. Shinagawa, Y. Kumagai and K. Urabe, *J. Porous Mater.,* 6(3), 185-190 (1999).
12. B.C. Jackson and G. Shawhan, IEEE Int. Symp. Electromagnetic Compatibility, Vol. 1, pp. 567-572 (1998).
13. R. Kumar, A. Kumar and D. Kumar, Proc. Int. Conf. Electromagnetic Interference and Compatibility, pp. 447-450 (1997).
14. L.G. Bhatgadde and S. Joseph, Proc. Int. Conf. Electromagnetic Interference and Compatibility, pp. 443-445 (1997).
15. A. Sidhu, J. Reike, U. Michelsen, R. Messinger, E. Habiger and J. Wolf, IEEE Int. Symp. Electromagnetic Compatibility, pp. 102-105 (1997).
16. J. Hajdu, *Trans. Inst. Metal Finish.,* 75(Pt. 1), B7-B10 (1997).
17. G. Klemmer, Special Areas Annual Technical Conf. — ANTEC, Conf. Proc., Vol. 3, pp. 3430-3432 (1996).
18. D. Gwinner, P. Scheyrer and W. Fernandez, Proc., Annual Technical Conf. — Soc. Vacuum Coaters, p. 336 (1996).
19. M.S. Bhatia, Proc. 4th Int. Conf. Electromagnetic Interference and Compatibility, pp. 321-324 (1995).
20. L. Zhang, W. Li, J. Liu and B. Ren, *Cailiao Gongcheng/J. Mater. Eng.,* No. 7, pp. 38-41 (1995).
21. N.V. Mandich, *Plat. Surf. Finish.,* 81(10), 60-63 (1994).
22. B.C. Jackson and P. Kuzyk, IEE Conf. Publication, No. 396, pp. 119-124 (1994).
23. C. Nagasawa, Y. Kumagai, K. Urabe and S. Shinagawa, *J. Porous Mater.,* 6(3), 247-254 (1999).

24. D.S. Dixon and J. Masi, IEEE Int. Symp. Electromagnetic Compatibility, Vol. 2, pp. 1035-1040 (1998).
25. P.J.D. Mason, Proc. Annual Technical Conf. — Soc. Vacuum Coaters, pp. 192-197 (1994).
26. C.A. Grimes, IEEE Aerospace Applications Conf. Proc., pp. 211-221 (1994).
27. W.J. Biter, P.J. Jamnicky and W. Coburn, Int. SAMPE Electronics Conf., Vol. 7, pp. 234-242 (1994).
28. L. Xing, J. Liu and S. Ren, *Cailiao Gongcheng/J. Mater. Eng.*, No. 1, pp. 19-21 (1998).
29. L. Rupprecht and C. Hawkinson, Medical Device & Diagnostic Industry 21(1), (1999).
30. S. Tan, M. Zhang and H. Zeng, *Cailiao Gongcheng/J. Mater. Eng.*, No. 5, pp. 6-9 (1998).
31. R. Charbonneau, Proc. Int. SAMPE Symp. Exhib., Vol. 43, No. 1, pp. 833-844 (1998).
32. J.M. Kolyer, Proc. Int. SAMPE Symp. Exhib., Vol. 43, No. 1, pp. 810-822 (1998).
33. S.L. Thompson, *Evaluation Eng.,* 37(7), 62-63, 65 (1998).
34. D.A. Olivero and D.W. Radford, *J. Reinforced Plast. Composites,* 17(8), 674-690 (1998).
35. K.P. Sau, T.K. Chaki, A. Chakraborty and D. Khastgir, *Plast. Rubber Composites Process. Appl.,* 26(7), 291-297 (1997).
36. W.B. Genetti, B.P. Grady and E.A. O'Rear, Electronic Packaging Materials Science IX, Materials Research Society Symp. Proc., Vol. 445, pp. 153-158 (1997).
37. J. Mao, J. Chen, M. Tu, W. Huang and Y. Liu, *Gongcheng Cailiao/J. Functional Mater.,* 28(2), 137-139 (1997).
38. L. Rupprecht, Proc. Conf. Plastics for Portable and Wireless Electronics, pp. 12-20 (1996).
39. B.K. Bachman, Proc. Conf. Plastics for Portable and Wireless Electronics, pp. 7-11 (1996).
40. S. Schneider, *Kunststoffe Plast. Europe,* 87(4), 487-488 (1997).
41. S. Schneider, *Kunststoffe Plast. Europe,* 87(4), 26 (1997).
42. Ch-Y. Huang and J-F. Pai, *J. Appl. Polym. Sci.,* 63(1), 115-123 (1997).
43. M.W.K. Rosenow and J.A.E. Bell, Materials Annual Tech. Conf. – ANTEC, Conf. Proc., Vol. 2, pp. 1492-1498 (1997).
44. M.W.K. Rosenow and J.A.E. Bell, Proc. Int. SAMPE Symp. Exhib., Vol. 43, No. 1, pp. 854-864 (1998).
45. M.A. Saltzberg, A.L. Neller, C.S. Harvey, T.E. Borninski and R.J. Gordon, *Circuit World,* 22(3), 67-68 (1996).
46. J. Wang, V.V. Varadan and V.K. Varadan, *SAMPE J.,* 32(6), 18-22 (1996).
47. J.V. Masi and D.S. Dixon, Int. SAMPE Electronics Conf., Vol. 7, pp. 243-251 (1994).
48. H. Rahman, J. Dowling and P.K. Saha, *J. Mater. Proc. Tech.,* 54(1-4), 21-28 (1995).
49. A.A. Dani and A.A. Ogale, 50 Years of Progress in Materials and Science Technology, Int. SAMPE Technical Conf., Vol. 26, pp. 689-699 (1994).
50. D.W. Radford, *J. Adv. Mater.,* 26(1), 45-53 (1994).
51. C.M. Ma, A.T. Hu and D.K. Chen, *Polym. Polym. Composites,* 1(2), 93-99 (1993).
52. K. Miyashita and Y. Imai, *Int. Prog. Urethanes,* 6, 195-218 (1993).
53. L. Li, P. Yih and D.D.L. Chung, *J. Electron. Mater.,* 21(11), 1065-1071 (1992).
54. L. Li and D.D.L. Chung, *Composites,* 25(3), 215-224 (1994).
55. L. Li and D.D.L. Chung, *Polym. Composites,* 14(5), 361-366 (1993).
56. L. Li and D.D.L. Chung, *Polym. Composites,* 14(6), 467-472 (1993).
57. L. Li and D.D.L. Chung, *Composites,* 22(3), 211-218 (1991).

58. M. Zhu and D.D.L. Chung, *J. Electron. Packag.,* 113, 417-420 (1991).

59. M. Zhu and D.D.L. Chung, *Composites,* 23(5), 355-364 (1992).

60. J.A. Pomposo, J. Rodriguez and H. Grande, *Synth. Met.,* 104(2), 107-111 (1999).

61. J-S. Park, S-H. Ryn and O-H. Chung, Annual Technical Conf. — ANTEC, Conf. Proc., Vol. 2, pp. 2410-2414 (1998).

62. S. Courric and V.H. Tran, *Polymer,* 39(12), 2399-2408 (1998).

63. M. Angelopoulos, Proc. Conf. Plastics for Portable and Wireless Electronics, p. 66 (1997).

64. T. Makela, S. Pienimaa, T. Taka, S. Jussila and H. Isotalo, *Synth. Met.,* 85(1-3), 1335-1336 (1997).

65. R.S. Kohlman, Y.G. Min, A.G. MacDiarmid and A.J. Epstein, *J. Eng. Appl. Sci.,* 2, 1412-1416 (1996).

66. K. Naishadham, Int. SAMPE Electronics Conf., Vol. 7, pp. 252-265 (1994).

67. A. Kaynak, A. Polat and U. Yilmazer, *Mater. Res. Bull.,* 31(10), 1195-1206 (1996).

68. A. Kaynak, *Mater. Res. Bull.,* 31(7), 845-860 (1996).

69. J. Joo, A.G. MacDiarmid and A.J. Epstein, Ann. Tech. Conf. — ANTEC, Conf. Proc., Vol. 2, pp. 1672-1677 (1995).

70. C.P.J.H. Borgmans and R.H. Glaser, *Evaluation Eng.,* 34(7), S-32–S-37 (1995).

71. P. Yan, *Scientific American,* 273(1), 82-87 (1995).

72. P.J. Mooney, *JOM,* 46(3), 44-45 (1994).

73. H.H. Kuhn, A.D. Child and W.C. Kimbrell, *Synth. Mater.,* 71(1-3), 2139-2142 (1995).

74. M.T. Nguyen and A.F. Diaz, *Adv. Mater.,* 6(11), 858-860 (1994).

75. J. Unsworth, C. Conn, Z. Jin, A. Kaynak, R. Ediriweera, P. Innis and N. Booth, *J. Intelligent Mater. Syst. Struct.,* 5(5), 595-604 (1994).

76. A. Kaynak, J. Unsworth, R. Clout, A.S. Mohan and G.E. Beard, *J. Appl. Polym. Sci.,* 54(3), 269-278 (1994).

77. A. Kaynak, A.S. Mohan, J. Unsworth and R. Clout, *J. Mater. Sci. Lett.,* 13(15), 1121-1123 (1994).

78. W. Sauerer, *Galvanotechnik,* 85(5), 1467-1472 (1994).

79. M.P. de Goefe and L.W. Steenbakkers, *Kunststoffe Plast. Eur.,* 84, 16-18 (1994).

80. X. Fu and D.D.L. Chung, *Carbon,* 36(4), 459-462 (1998).

81. L. Gnecco, *Evaluation Eng.,* 38(3), (1999).

82. S-S. Lin, *SAMPE J.,* 30(5), 39-45 (1994).

83. Y. Kurosaki and R. Satake, IEEE Int. Symp. Electromagnetic Compatibility, pp. 739-740 (1994).

84. X. Luo and D.D.L. Chung, *Composites,* Part B, 30(3), 227-231 (1999).

85. M. Zhu, Y. Qiu and J. Tian, *Cailiao Gongcheng/J. Functional Mater.,* 29(6), 645-647 (1998).

86. D.A. Case and M.J. Oliver, *Compliance Eng.,* 16(2), 40, 42, 44, 46, 48-49 (1999).

87. S. Hudak, *Evaluation Eng.,* 37(8), (1998).

88. B.N. Prakash and L.D. Roy, Proc. Int. Conf. Electromagnetic Interference and Compatibility, pp. 1-2 (1997).

89. S.H. Peng and K. Zhang, Annual Tech. Conf. — ANTEC, Conf. Proc., Vol. 2, pp. 1216-1218 (1998).

90. S.K. Das, J. Nuebel and B. Zand, IEEE Int. Symp. Electromagnetic Compatibility, pp. 66-71 (1997).

91. S.H. Peng and W.S.V. Tzeng, IEEE Int. Symp. Electromagnetic Compatibility, pp. 94-97 (1997).

92. P. O'Shea, *Evaluation Eng.,* 36(8), (1997).

93. P. O'Shea, *Evaluation Eng.,* 35(8), (1996).

94. Anonymous, *Electron. Eng.,* 68(834), (1996).
95. B. Lee, *Engineering,* 236(10), 32-33 (1995).
96. R.A. Rothenberg, D.C. Inman and Y. Itani, IEEE Int. Symp. Electromagnetic Compatibility, p. 818 (1994).
97. B.D. Mottahed and S. Manoochehri, *Polym. Eng. Sci.,* 37(3), 653-666 (1997).
98. X. Shui and D.D.L. Chung, *J. Electron. Packag.,* 119(4), 236-238 (1997).
99. X. Shui and D.D.L. Chung, *J. Electron. Mater.,* 26(8), 928-934 (1997).
100. X. Shui and D.D.L. Chung, *J. Electron. Mater.,* 24(2), 107-113 (1995).
101. Y. Ramadin, S.A. Jawad, S.M. Musameh, M. Ahmad, A.M. Zihlif, A. Paesano, E. Martuscelli and G. Ragosta, *Polym. Int.,* 34(2), 145-150 (1994).
102. M-S. Lin, IEEE Int. Symp. Electronmagnetic Compatibility, pp. 112-115 (1994).
103. H-K. Chiu, M-S. Lin and C.H. Chen, *IEEE Trans. Electromagnetic Compatibility,* 39(4), 332-339 (1997).
104. J.C. Roberts and P.D. Weinhold, *J. Composite Mater.,* 29(14), 1834-1849 (1995).
105. D.A. Olivero and D.W. Radford, *SAMPE J.,* 33(1), 51-57 (1997).
106. M. Choate and G. Broadbent, Thermosets: The True Engineering Polymers, Technical Papers, Regional Tech. Conf. — Soc. Plast. Eng., pp. 69-82 (1996).
107. P.D. Wienhold, D.S. Mehoke, J.C. Roberts, G.R. Seylar and D.L. Kirkbride, Int. SAMPE Tech. Conf., Vol. 30, pp. 243-255 (1998).
108. A. Fernyhough and Y. Yokota, *Materials World,* 5(4), 202-204 (1997).
109. T. Hiramoto, T. Terauchi and J. Tomibe, Electrical Overstress/Electrostatic Discharge Symp. Proc., ESD Assoc., pp. 18-21 (1998).
110. L.G. Morin Jr. and R.E. Duvall, Proc. Int. SAMPE Symp. Exhib., Vol. 43, No. 1, pp. 874-881 (1998).
111. J.R. Gaier and J. Terry, Int. SAMPE Electronics Conf., Vol. 7, pp. 221-233 (1994).
112. J.R. Gaier, M.L. Davidson and R.K. Shively, Technology Transfer in a Global Community, Int. SAMPE Tech. Conf., Vol. 28, pp. 1136-1147 (1996).
113. J.M. Liu, S.N. Vernon, A.D. Hellman and T.A. Campbell, Proc. SPIE — The Int. Soc. Optical Eng., Vol. 2459, pp. 60-68 (1995).
114. D.A. Olivero and D.W. Radford, Technology Transfer in a Global Community, Int. SAMPE Tech. Conf., Vol. 28, pp. 1110-1121 (1996).
115. X. Luo and D.D.L. Chung, *Carbon,* 34(10), 1293-1294 (1996).
116. K. Shibata, T. Okabe, K. Saito, T. Okayama, M. Shimada, A. Yamamura and R. Yamamoto, *J. Porous Mater.,* 4(4), 269-275 (1997).
117. T. Okabe, K. Saito, H. Togawa and Y. Kumagai, *Zairyo/J. Soc. Mater. Sci.,* Japan, 44(498), 288-291 (1995).
118. B.M. Rakshit, Proc. 1995 4th Int. Conf. Electromagnetic Interference Compatibility, IEEE, pp. 413-415 (1995).
119. C.R. Paul, *Introduction to Electromagnetic Compatibility,* K. Chang, Ed., John Wiley, New York, p. 649 (1992).

5 Cement-Based Electronics

SYNOPSIS Cement-based electronics in the form of electrical circuit elements (conductor and diode in the form of a pn-junction), sensors of strain and damage, thermistor, and thermoelectric device are reviewed. They enable a concrete structure to provide electronic functions.

RELEVANT APPENDICES: *A, B, C, E, F, I*

5.1 INTRODUCTION

Cement, such as Portland cement, has long been used as a structural material. It can also be used as an electronic material that provides electrical circuit elements, sensors, and thermoelectric devices. The use of a structural material for electronics enables a structure to provide electronic functions, making it multifunctional and smart.

Due to the large volume of usage of structural materials they must be inexpensive. In contrast, conventional electronic materials such as semiconductors are relatively expensive, particularly when vacuum processing, single crystals, and high purity are required. Using a structural material for electronics results in low-cost electronics.

A structural material must have good mechanical properties. Semiconductors tend to have poor mechanical properties. The use of a structural material for electronics results in electronics that are mechanically rugged.

This chapter is a review of cement-based electronics, an emerging field that is scientifically rich and technologically relevant.

5.2 BACKGROUND ON CEMENT-MATRIX COMPOSITES

Cement-matrix composites include concrete, mortar, and cement paste. They also include steel-reinforced concrete, i.e., concrete containing steel reinforcing bars. Other fillers, called admixtures, can be added to the mix to improve the properties of the composite. Admixtures are discontinuous, so they can be included in the mix. They can be particles, such as silica fume and latex; short fibers, such as polymer, steel, glass, or carbon fibers; or they can be liquids, such as methylcellulose aqueous solution, water-reducing agent, defoamer, etc. The focus of this chapter is admixtures for rendering composite electronic abilities while maintaining or even improving structural properties.

Cement-matrix composites for electronics include those containing short carbon fibers (for sensing strain, damage and temperature, for rendering p-type behavior, and for electromagnetic radiation reflection), short steel fibers (for thermoelectricity and for rendering n-type behavior), and silica fume (for helping fiber dispersion). This section provides background on cement-matrix composites, with emphasis on carbon fiber cement-matrix composites.

Carbon fiber cement-matrix composites are structural materials that are rapidly gaining in importance due to the decrease in carbon fiber cost[1] and the increasing demand for superior structural and functional properties. These composites contain short carbon fibers, typically 5 mm in length, as they can be used as an admixture in concrete (whereas continuous fibers cannot simply be added to the concrete mix). Short fibers are also less expensive than continuous fibers. However, due to the weak bond between carbon fibers and the cement matrix, continuous fibers[2-4] are much more effective in reinforcing concrete. Surface treatment of carbon fibers (by heating[5] or by using ozone,[6,7] silane,[8] SiO_2 particles,[9] or hot NaOH solution[10]) is useful for improving the bond between fiber and matrix, thereby improving the properties of the composite. In the case of surface treatment by ozone or silane, the improved bond is due to the enhanced wettability by water. Admixtures such as latex,[6,11] methyscellulose,[6] and silica fume[12] also help the bond.

The effect of carbon fiber addition on the properties of concrete increases with fiber volume fraction,[13] unless the fiber volume fraction is so high that the air void content becomes excessively high.[14] (The air void content increases with fiber content, and air voids tend to have a negative effect on many properties, such as the compressive strength.) In addition, the workability of the mix decreases with fiber content.[13] Moreover, the cost increases with fiber content. Therefore, a rather low volume fraction of fibers is desirable. A fiber content as low as 0.2 vol. % is effective,[15] although fiber contents exceeding 1 vol. % are more common.[16-20] The required fiber content increases with the particle size of the aggregate, as the flexural strength decreases with increasing particle size.[21]

Effective use of carbon fibers in concrete requires dispersion of the fibers in the mix. The dispersion is enhanced by using silica fume as an admixture.[14,22-24] A typical silica fume content is 15% by weight of cement.[14] The silica fume is typically used along with a small amount (0.4% by weight of cement) of methylcellulose for helping the dispersion of the fibers and the workability of the mix.[14] Latex (typically 15–20% by weight of cement) is much less effective for helping the fiber dispersion, but it enhances workability, flexural strength, flexural toughness, impact resistance, frost resistance, and acid resistance.[14,25,26] The ease of dispersion increases with decreasing fiber length.[24]

The improved structural properties rendered by carbon fiber addition pertain to increased tensile and flexible strengths, increased tensile ductility and flexural toughness, enhanced impact resistance, reduced drying shrinkage, and improved freeze-thaw durability.[13-15,17-25,27-38] Tensile and flexural strengths decrease with increasing specimen size such that the size effect becomes larger as the fiber length increases.[39] The low drying shrinkage is valuable for large structures and for use in repair,[40,41] and in joining bricks in a structure.[42,43]

The functional properties rendered by carbon fiber addition pertain to strain-sensing ability,[7,44-60] temperature-sensing ability,[61-68] damage-sensing ability,[44,48,69-71] thermoelectric behavior,[63-68] thermal insulation ability,[72-74] electrical conduction ability[75-85] (to facilitate cathodic protection of embedded steel and to provide electrical grounding or connection), and radio wave reflection/adsorption ability[86-90] (for EMI shielding, for lateral guidance in automatic highways, and for television image transmission).

In relation to the structural properties, carbon fibers compete with glass, polymer, and steel fibers.[18,27-29,32,36-38,91] Carbon fibers (isotropic pitch-based)[1,91] are advantageous in their superior ability to increase the tensile strength of concrete, even though the tensile strength, modulus, and ductility of isotropic pitch-based carbon fibers are low compared to most other fibers. Carbon fibers are also advantageous for relative chemical inertness.[92] PAN-based carbon fibers are also used,[17,19,22,33] although they are more commonly used as continuous fibers. Carbon-coated glass fibers[93,94] and submicron-diameter carbon filaments[86,87] are even less commonly used, although the former are attractive for the low cost of glass fibers and the latter for high radio wave reflectivity. C-shaped carbon fibers are more effective for strengthening than are round carbon fibers,[95] but their relatively large diameter makes them less attractive. Carbon fibers can be used in concrete together with steel fibers, as the addition of short carbon fibers to steel fiber-reinforced mortar increases the fracture toughness of the interfacial zone between steel fibers and the cement matrix.[96] Carbon fibers can also be used in concrete with steel bars,[97,98] or with carbon fiber-reinforced polymer rods.[99]

In relation to most functional properties, carbon fibers are exceptional compared to other fiber types. Carbon fibers are electrically conducting, in contrast to glass and polymer fibers, which are not. Steel fibers are conducting, but their typical diameters (≥ 60 μm) are much larger than the diameter of a typical carbon fiber (15 μm). The combination of electrical conductivity and small diameter makes carbon fibers superior to the other fiber types in the area of strain sensing and electrical conduction. However, carbon fibers are inferior to steel fibers for providing thermoelectric composites because of the high electron concentration in steel and the low hole concentration in carbon.

Although carbon fibers are thermally conducting, addition of carbon fibers to concrete lowers thermal conductivity,[72] thus allowing applications related to thermal insulation. This effect of carbon fiber addition is due to the increase in air void content. The electrical conductivity of carbon fibers is higher than that of the cement matrix by about eight orders of magnitude, whereas the thermal conductivity of carbon fibers is higher than that of the cement matrix by only one or two orders of magnitude. As a result, electrical conductivity is increased upon carbon fiber addition despite the increase in air void content, but thermal conductivity is decreased.

The use of pressure after casting,[100] and extrusion[101,102] can result in composites with superior microstructure and properties. Moreover, extrusion improves the shapability.[102]

5.3 CEMENT-BASED ELECTRICAL CIRCUIT ELEMENTS

5.3.1 CONDUCTOR

Short carbon fibers are particularly effective for enhancing the electrical conductivity of cement.[80] Figure 5.1[80] illustrates the volume electrical resistivity of composites at seven days of curing. The resistivity decreases with increasing fiber volume fraction, whether or not a second filler (silica fume or sand) is present. When sand is absent, the addition of silica fume decreases the resistivity at all carbon fiber volume fractions except the highest of 4.24%; the decrease is most significant at the lowest fiber volume fraction of 0.53%. When sand is present, the addition of silica fume similarly decreases resistivity such that the decrease is most significant at fiber volume fractions below 1%. When silica fume is absent, the addition of sand decreases resistivity only when the fiber volume fraction is below about 0.5%; at high fiber volume fractions, the addition of sand increases resistivity due to the porosity induced by the sand. Thus, the addition of a second filler that is essentially nonconducting decreases the resistivity of the composite only at low volume fractions of the carbon fibers, and the maximum fiber volume fraction for the resistivity to decrease is larger when the particle size of the filler is smaller. The resistivity decrease is attributed to improved fiber dispersion because of the presence of the second filler. Consistent with improved fiber dispersion is increased flexural toughness and strength.

5.3.2 DIODE

The pn-junction is the junction between a p-type conductor (a conductor with holes as the majority carrier) and an n-type conductor (a conductor with electrons as the

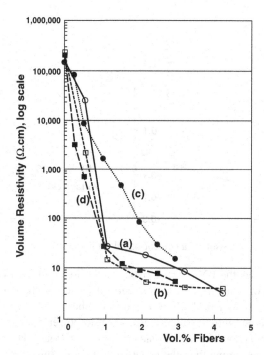

FIGURE 5.1 Variation of the volume electrical resistivity with carbon fiber volume fraction: (**a**) without sand, with methylcellulose, without silica fume; (**b**) without sand, with methylcellulose, with silica fume; (**c**) with sand, with methyscellulose, without silica fume; (**d**) with sand, with methyscellulose, with silica fume.

majority carrier). The pn-junction is ideally rectifying, i.e., the current-voltage characteristic is such that the current is large when the applied voltage is positive on the p-side relative to the n-side, and is small when the applied voltage is positive on the n-side relative to the p-side. The pn-junction is an electronic device that is central to electrical circuitry because of its importance to diodes and transistors. Akin to the pn-junction is the n-n$^+$ junction, a junction between a weakly n-type conductor and a strongly n-type (i.e., n$^+$) conductor.

In the field of electrical engineering, a pn-junction is obtained by allowing a p-type semiconductor to contact an n-type semiconductor. These semiconductors are obtained by doping with appropriate impurities that serve as electron donors (for n-type semiconductors) or electron acceptors (for p-type semiconductors).

Cement is inherently n-type, although it is only slightly n-type.[65] Upon addition of a sufficient amount of short carbon fibers to cement, a composite that is p-type is obtained.[63-66] Upon addition of short steel fibers to cement, a composite that is strongly n-type is obtained.[103] In other words, the cement matrix contributes to conduction by electrons; carbon fibers contribute to conduction by holes; and steel fibers contribute to conduction by electrons.

Electric current rectification has been observed in a cement junction, as obtained by the separate pouring of electrically dissimilar cement mixes side by side.[104]

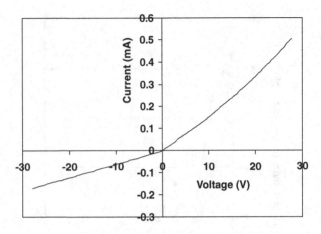

FIGURE 5.2 I-V characteristic of a cement-based pn junction.

Figure 5.2 shows the I-V characteristic for the junction involving carbon-fiber (1.0% by weight of cement), silica-fume cement paste (p-type, with absolute thermoelectric power -0.48 ± 0.11 µV/°C), and steel fiber (0.5% by weight of cement) cement paste (n-type, with absolute thermoelectric power 53.3 ± 4.8 µV/°C). This pn-junction is rectifying. The I-V characteristic is nonlinear. The slope is steeper in the positive voltage regime than the negative voltage regime.

The asymmetry in the I-V characteristic on the two sides of the origin for various junctions is in sharp contrast to the symmetry for a homogeneous piece of cement paste without a junction.[61] Hence, the asymmetry is attributed to the junction itself.

The junctions are rectifying; the magnitude of current is much larger when the voltage is positive than when the voltage is negative. The rectification is attributed to the asymmetric electron flow resulting from the junction. Due to the high concentration of electrons on the n-side, electrons flow by diffusion from the n-side to the p-side across the junction. This flow is enhanced by a positive voltage that lowers the contact potential at the junction. When the voltage is negative, the contact potential is high, causing the diffusion current to be low. However, electrons are swept from the p-side to the n-side under the electric field associated with the high contact potential at the junction, resulting in a drift current in the direction opposite to the diffusion current. The drift current is enhanced by a more negative voltage, but is low due to the low electron concentration on the p-side.

5.4 CEMENT-BASED SENSORS

5.4.1 STRAIN SENSOR

The electrical resistance of strain-sensing concrete changes reversibly with strain such that the gage factor (fractional change in resistance per unit strain) is up to 700 under compression or tension.[7,44-60] The resistance (DC/AC) increases reversibly upon tension and decreases reversibly upon compression because of fiber pull-out upon microcrack opening (< 1 µm), and the consequent increase in fiber-matrix contact resistivity. The concrete contains as low as 0.2 vol.% short carbon fibers, preferably those that have been surface treated. The fibers do not need to touch one another in the composite. The treatment improves the wettability with water. The presence of a large aggregate decreases the gage factor, but the strain-sensing ability remains sufficient for practical use. Strain-sensing concrete works even when data acquisition is wireless. Applications include structural vibration control and traffic monitoring.

Figure 5.3(a) shows the fractional change in resistivity along the stress axis, and the strain during repeated compressive loading at an increasing stress amplitude for carbon-fiber latex cement paste at 28 days of curing. Figure 5.3(b) shows the corresponding variation of stress and strain during the repeated loading. The strain varies linearly with the stress up to the highest stress amplitude (Figure 5.3(b)). The strain returns to zero at the end of each cycle of loading. The resistivity decreases upon loading in every cycle (due to fiber push-in) and increases upon unloading in every cycle (due to fiber pull-out). The resistivity has a net increase after the first cycle because of damage. Little further damage occurs in subsequent cycles, as shown by the resistivity after unloading not increasing much after the first cycle. The greater the strain amplitude, the more the resistivity decrease during loading, although resistivity and strain are not linearly related. The effects in the figures were similarly observed in carbon-fiber silica-fume cement paste at 28 days of curing.

Figures 5.4 and 5.5 illustrate the fractional changes in the longitudinal and transverse resistivities, respectively, for carbon-fiber silica-fume cement paste at 28 days of curing during repeated unaxial tensile loading at increasing strain amplitudes. The strain essentially returns to zero at the end of each cycle, indicating elastic deformation. The longitudinal strain is positive (i.e., elongation); the transverse strain is negative (i.e., shrinkage due to the Poisson effect). Both longitudinal and transverse resistivities increase reversibly upon uniaxial tension. The reversibility of both strain and resistivity is more complete in the longitudinal direction than in the transverse direction. The gage factor is 89 and −59 for the longitudinal and transverse resistances, respectively.

Figures 5.6 and 5.7 show corresponding results for silica-fume cement paste. The strain is essentially totally reversible in both the longitudinal and transverse directions, but the resistivity is only partly reversible in both directions, in contrast to the reversibility of the resistivity when fibers are present (Figures 5.4 and 5.5). As in the case with fibers, both longitudinal and transverse resistivities increase upon uniaxial tension. However, the gage factor is only 7.2 and −7.1 for Figures 5.6 and 5.7, respectively.

Comparison of Figures 5.4 and 5.5 (with fibers) with Figures 5.6 and 5.7 (without fibers) shows that fibers greatly enhance the magnitude and reversibility of the resistivity effect. The gage factors are much smaller in magnitude when fibers are absent.

The increase in both longitudinal and transverse resistivities upon uniaxial tension for cement pastes, whether with or without fibers, is attributed to defect (e.g., microcrack) generation. In the presence of fibers, fiber bridging across microcracks occurs, and slight fiber pull-out takes place upon tension, enhancing the possibility of microcracks closing and causing more reversibility in the resistivity change. The

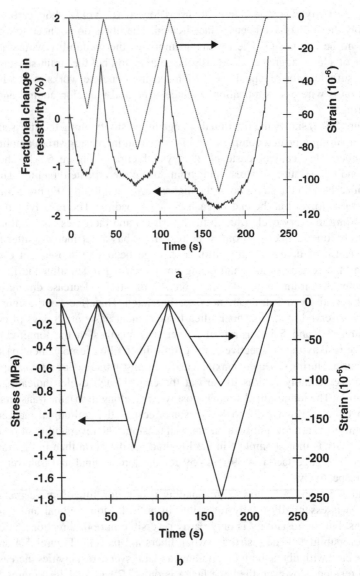

FIGURE 5.3 Variation of the fractional change in volume electrical resistivity with time (a), of the stress with time (b), and of the strain (negative for compressive strain) with time (a,b) during dynamic compressive loading at increasing stress amplitudes within the elastic regime for carbon-fiber latex cement paste at 28 days of curing.

fibers are much more electrically conductive than the cement matrix. The presence of the fibers introduces interfaces between fibers and matrix. The degradation of the fiber-matrix interface, due to fiber pull-out or other mechanisms, is an additional type of defect generation that will increase the resistivity of the composite. Therefore, the presence of fibers greatly increases the gage factor.

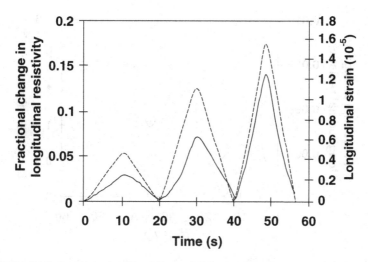

FIGURE 5.4 Variation of the fractional change in longitudinal electrical resistivity with time (solid curve) and of the strain with time (dashed curve) during dynamic uniaxial tensile loading at increasing stress amplitudes within the elastic regime for carbon-fiber silica-fume cement paste.

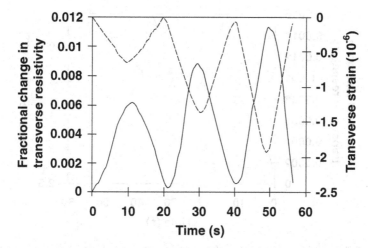

FIGURE 5.5 Variation of the fractional change in transverse electrical resistivity with time (solid curve) and of the strain with time (dashed curve) during dynamic uniaxial tensile loading at increasing stress amplitudes within the elastic regime for carbon-fiber silica-fume cement paste.

Transverse resistivity increases upon uniaxial tension, even though the Poisson effect causes the transverse strain to be negative. This means that the effect of the transverse resistivity increase overshadows the effect of transverse shrinkage. The resistivity increase is a consequence of uniaxial tension. In contrast, under uniaxial compression, resistance in the stress direction decreases at 28 days of curing. Hence,

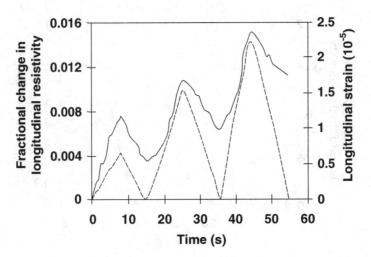

FIGURE 5.6 Variation of the fractional change in longitudinal electrical resistivity with time (solid curve) and of the strain with time (dashed curve) during dynamic uniaxial tensile loading at increasing stress amplitudes within the elastic regime for silica-fume cement paste.

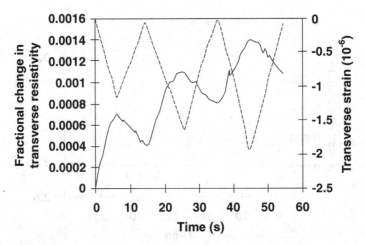

FIGURE 5.7 Variation of the fractional change in transverse electrical resistivity with time (solid curve) and of the strain with time (dashed curve) during dynamic uniaxial tensile loading at increasing stress amplitudes within the elastic regime for silica-fume cement paste.

the effect of uniaxial tension on transverse resistivity, and that of uniaxial compression on longitudinal resistivity are different; the gage factors are negative and positive, respectively.

The similarity of the resistivity change in longitudinal and transverse directions under uniaxial tension suggests similarity for other directions as well. This means the resistance can be measured in any direction to sense the occurrence of tensile loading. Although the gage factor is comparable in both longitudinal and transverse directions, the fractional change in resistance under uniaxial tension is much higher in the longitudinal direction than in the transverse direction. Thus, the use of longitudinal resistance for practical self-sensing is preferred.

5.4.2 DAMAGE SENSOR

Concrete, with or without admixtures, is capable of sensing major and minor damage — even damage during elastic deformation — due to the electrical resistivity increase that accompanies damage.[44,48,69-71] That both strain and damage can be sensed simultaneously through resistance measurement means that the strain/stress condition (during dynamic loading) under which damage occurs can be obtained, thus facilitating damage origin identification. Damage is indicated by a resistance increase, which is larger and less reversible when the stress amplitude is higher. The resistance increase can be a sudden increase during loading. It can also be a gradual shift of baseline resistance.

Figure 5.8(a)[70] shows the fractional change in resistivity along the stress axis, as well as the strain during repeated compressive loading at an increasing stress amplitude, for plain cement paste at 28 days of curing. Figure 5.8(b) shows the corresponding variation of stress and strain during the repeated loading. The strain varies linearly with the stress up to the highest stress amplitude (Figure 5.8(b)). The strain returns to zero at the end of each cycle of loading. During the first loading, resistivity increases due to damage initiation. During the subsequent unloading, resistivity continues to increase, probably due to the opening of the microcracks generated during loading. During the second loading, resistivity decreases slightly as the stress increases up to the maximum stress of the first cycle (probably due to closing of the microcracks), and then increases as the stress increases beyond this value (probably due to the generation of additional microcracks). During unloading in the second cycle, resistivity increases significantly (probably due to opening of the microcracks). During the third loading, resistivity essentially does not change (or decreases very slightly) as the stress increases to the maximum stress of the third cycle (probably due to the balance between microcrack generation and microcrack closing). Subsequent unloading causes resistivity to increase very significantly (probably due to opening of the microcracks).

Figure 5.9 shows the fractional change in resistance, strain, and stress during repeated compressive loading at increasing and decreasing stress amplitudes for carbon fiber (0.18 vol.%) concrete at 28 days of curing. The highest stress amplitude is 60% of the compressive strength. A group of cycles in which the stress amplitude increases cycle by cycle, and then decreases cycle by cycle to the initial low stress amplitude, is hereby referred to as a group. Figure 5.9 shows the results for three groups. The strain returns to zero at the end of each cycle for the stress amplitudes, indicating elastic behavior. The resistance decreases upon loading in each cycle, as in Figure 5.3. An extra peak at the maximum stress of a cycle grows as the stress amplitude increases, resulting in two peaks per cycle. The original peak (strain induced) occurs at zero stress, while the extra peak (damage induced) occurs at the maximum stress. Hence, during loading from zero stress within a cycle, the resistance

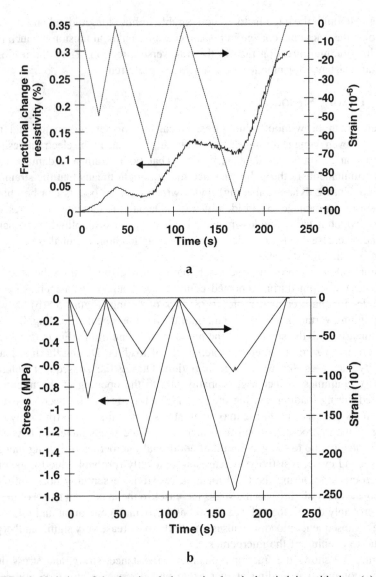

FIGURE 5.8 Variation of the fractional change in electrical resistivity with time (**a**), of the stress with time (**b**), and of the strain (negative for compressive strain) with time (**a,b**) during dynamic compressive loading at increasing stress amplitudes within the elastic regime for silica-fume cement paste at 28 days of curing.

drops and then increases sharply, reaching the maximum resistance of the extra peak at the maximum stress of the cycle. Upon subsequent unloading, resistance decreases and then increases as unloading continues, reaching the maximum resistance of the original peak at zero stress. In the part of this group where the stress amplitude decreases cycle by cycle, the extra peak diminishes and disappears, leaving the original peak as the sole peak. In the part of the second group where the stress

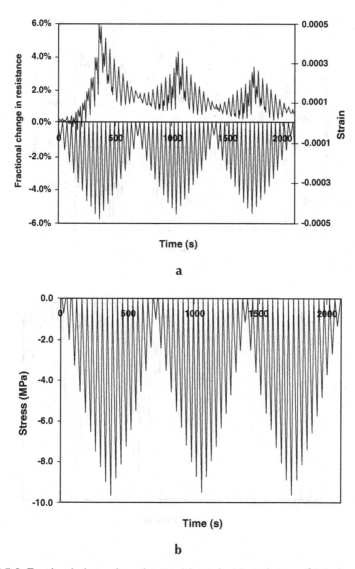

FIGURE 5.9 Fractional change in resistance (**a**), strain (**a**), and stress (**b**) during repeated compressive loading at increasing and decreasing stress amplitudes, the highest of which was 60% of the compressive strength, for carbon fiber concrete at 28 days of curing.

amplitude increases cycle by cycle, the original peak is the sole peak, except that the extra peak returns in a minor way (more minor than in the first group) as the stress amplitude increases. The extra peak grows as the stress amplitude increases, but, in the part of the second group in which the stress amplitude decreases cycle by cycle, it quickly diminishes and vanishes, as in the first group. Within each group, the amplitude of resistance variation increases as the stress amplitude increases, and decreases as the stress amplitude decreases.

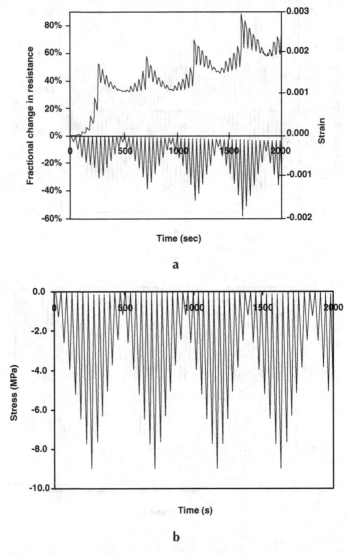

FIGURE 5.10 Fractional change in resistance (**a**), strain (**a**), and stress (**b**) during repeated compressive loading at increasing and decreasing stress amplitudes, the highest of which was >90% of the compressive strength, for carbon fiber concrete at 28 days of curing.

The greater the stress amplitude, the larger and less reversible the damage-induced resistance increase. If the stress amplitude has been experienced before, the damage-induced resistance increase is small, as shown by comparing the result of the second group with that of the first (Figure 5.9), unless the extent of damage is large (Figure 5.10 shows a highest stress amplitude of >90% the compressive strength). When the damage is extensive (as shown by a modulus decrease), damage-induced resistance increase occurs in every cycle, even at a decreasing stress ampli-

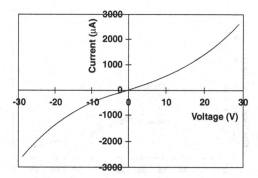

FIGURE 5.11 Current-voltage characteristic of carbon-fiber silica-fume cement paste at 38°C during stepped heating.

tude, and it can overshadow the strain-induced resistance decrease (Figure 5.10). Hence, damage-induced resistance increase occurs mainly during loading (even within the elastic regime), particularly at a stress above that in prior cycles, unless the stress amplitude is high and/or damage is extensive.

At a high stress amplitude, damage-induced resistance increases cycle by cycle as the increases in stress amplitude cause the baseline resistance to increase irreversibly (Figure 5.10). The baseline resistance in the regime of major damage (with a decrease in modulus) provides a measure of the extent of damage. This measure works in the loaded or unloaded state. In contrast, the measure using the damage-induced resistance increase (Figure 5.9) works only during stress increase and indicates the occurrence of damage and its extent.

5.4.3 THERMISTOR

A thermistor is a thermometric device consisting of a material (typically a simiconductor, but in this case a cement paste) whose electrical resistivity decreases with rise in temperature. The carbon fiber concrete described above for strain sensing is a thermistor due to its resistivity decreasing reversibly with increasing temperature;[61] its sensitivity is comparable to that of semiconductor thermistors. (The effect of temperature will need to be compensated by using the concrete as a strain sensor.) Without fibers, sensitivity is much lower.[61,62]

Figure 5.11[61] shows the current-voltage characteristic of carbon-fiber (0.5% by weight of cement) silica-fume (15% by weight of cement) cement paste at 38°C during stepped heating. The characteristic is linear below 5 V, and deviates positively from linearity beyond 5 V. The resistivity is obtained from the slope of the linear portion. The voltage at which the characteristic starts to deviate from linearity is referred to as the "critical voltage."

Figure 5.12 shows a plot of resistivity vs. temperature during heating and cooling for carbon-fiber silica-fume cement paste. The resistivity decreases upon heating, and the effect is quite reversible upon cooling. That the resistivity is slightly increased after a heating-cooling cycle is probably due to thermal degradation of the material. Figure 5.13 shows the Arrhenius plot of log conductivity (conductivity = 1/resistiv-

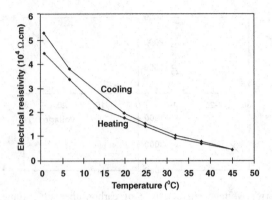

FIGURE 5.12 Plot of volume electrical resistivity vs. temperature during heating and cooling for carbon-fiber silica-fume cement paste.

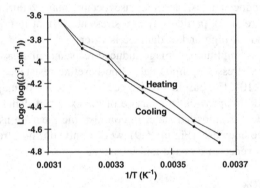

FIGURE 5.13 Arrhenius plot of log electrical conductivity vs. reciprocal absolute temperature for carbon-fiber silica-fume cement paste.

ity) vs. reciprocal absolute temperature. The slope of the plot gives the activation energy, which is 0.390 ± 0.014 and 0.412 ± 0.017 eV during heating and cooling, respectively.

Results similar to those of carbon-fiber silica-fume cement paste were obtained with carbon-fiber (0.5% by weight of cement) latex (20% by weight of cement) cement paste, silica-fume cement paste, latex cement paste, and plain cement paste. However, for all these four types of cement paste, (1) the resistivity is higher by about an order of magnitude, and (2) the activation energy is lower by about an order of magnitude, as shown in Table 5.1. The critical voltage is higher when fibers are absent.

5.5 CEMENT-BASED THERMOELECTRIC DEVICE

The Seebeck[63-68,103] effect is a thermoelectric effect that is the basis for thermocouples for temperature measurement. This effect involves charge carriers moving from a hot point to a cold point within a material, resulting in a voltage difference between

TABLE 5.1

Resistivity, Critical Voltage, and Activation Energy of Five Types of Cement Paste

Formulation	Resistivity at 20°C (Ω.cm)	Critical Voltage at 20°C (V)	Activation Energy (eV)	
			Heating	Cooling
Plain	$(4.87 \pm 0.37) \times 10^5$	10.80 ± 0.45	0.040 ± 0.006	0.122 ± 0.006
Silica fume	$(6.12 \pm 0.15) \times 10^5$	11.60 ± 0.37	0.035 ± 0.003	0.084 ± 0.004
Carbon fibers + silica fume	$(1.73 \pm 0.08) \times 10^4$	8.15 ± 0.34	0.390 ± 0.014	0.412 ± 0.017
Latex	$(6.99 \pm 0.12) \times 10^5$	11.80 ± 0.31	0.017 ± 0.001	0.025 ± 0.002
Carbon fibers + latex	$(9.64 \pm 0.08) \times 10^4$	8.76 ± 0.35	0.018 ± 0.001	0.027 ± 0.002

the two points. The Seebeck coefficient is the negative of the voltage difference per unit temperature difference between the two points. Negative carriers (electrons) make it more negative, and positive carriers (holes) make it more positive.

The Seebeck effect in carbon fiber-reinforced cement paste involves electrons and/or ions from the cement matrix[65] and holes from the fibers,[63-66] such that the two contributions are equal at the percolation threshold, a fiber content between 0.5% and 1.0% by weight of cement.[65] The hole contribution increases monotonically with increasing fiber content below and above the percolation threshold.[65]

Due to the free electrons in a metal, cement containing metal fibers such as steel fibers is even more negative in the thermoelectric power than cement without fiber.[103] The attainment of a very negative thermoelectric power is attractive since a material with a negative thermoelectric power and one with positive thermoelectric power are two very dissimilar materials, the junction of which is a thermocouple junction. (The greater the dissimilarity, the more sensitive the thermocouple.)

Table 5.2 and Figure 5.14 show the thermopower results. The absolute thermoelectric power is much more negative for all the steel-fiber cement pastes compared to all the carbon-fiber cement pastes. An increase of the steel fiber content from 0.5% to 1.0% by weight of cement makes the absolute thermoelectric power more negative, whether or not silica fume (or latex) is present. An increase of the steel fiber content also increases the reversibility and linearity of the change in Seebeck voltage with the temperature difference between the hot and cold ends, as shown by comparing the values of the Seebeck coefficient obtained during heating and cooling in Table 5.2. The values obtained during heating and cooling are close for the pastes with the higher steel fiber content, but are not so close for those with the lower steel fiber content. In contrast, for pastes with carbon fibers in place of steel fibers, the change in Seebeck voltage with the temperature difference is highly reversible for both carbon fiber contents of 0.5% and 1.0% by weight of cement, as shown in Table 5.2 by comparing the values of the Seebeck coefficient obtained during heating and cooling.

TABLE 5.2
Volume Electical Resistivity, Seebeck Coefficient (μV/°C) with Copper as the Reference, and the Absolute Thermoelectric Power (μV/°C) of Various Cement Pastes with Steel Fibers (S_f) or Carbon Fibers (C_f)

Cement Paste	Volume Fraction Fibers	Resistivity (Ω.cm)	Heating		Cooling	
			Seebeck Coefficient	Absolute Thermoelectric Power	Seebeck Coefficient	Absolute Thermoelectric Power
Plain	0	$(4.7 \pm 0.4) \times 10^5$	$+0.35 \pm 0.03$	-1.99 ± 0.03	$+0.38 \pm 0.05$	-1.96 ± 0.05
SF	0	$(5.8 \pm 0.4) \times 10^5$	$+0.31 \pm 0.02$	-2.03 ± 0.02	$+0.36 \pm 0.03$	-1.98 ± 0.03
L	0	$(6.8 \pm 0.6) \times 10^5$	$+0.28 \pm 0.02$	-2.06 ± 0.02	$+0.30 \pm 0.02$	-2.04 ± 0.02
S_f (0.5*)	0.10%	$(7.8 \pm 0.5) \times 10^4$	-51.0 ± 4.8	-53.3 ± 4.8	-45.3 ± 4.4	-47.6 ± 4.4
S_f (1.0*)	0.20%	$(4.8 \pm 0.4) \times 10^4$	-56.8 ± 5.2	-59.1 ± 5.2	-53.7 ± 4.9	-56.0 ± 4.9
S_f (0.5*) + SF	0.10%	$(5.6 \pm 0.5) \times 10^4$	-54.8 ± 3.9	-57.1 ± 3.9	-52.9 ± 4.1	-55.2 ± 4.1
S_f (1.0*) + SF	0.20%	$(3.2 \pm 0.3) \times 10^4$	-66.2 ± 4.5	-68.5 ± 4.5	-65.6 ± 4.4	-67.9 ± 4.4
S_f (0.5*) + L	0.085%	$(1.4 \pm 0.1) \times 10^5$	-48.1 ± 3.2	-50.4 ± 3.2	-45.4 ± 2.9	-47.7 ± 2.9
S_f (1.0*) + L	0.17%	$(1.1 \pm 0.1) \times 10^5$	-55.4 ± 5.0	-57.7 ± 5.0	-54.2 ± 4.5	-56.5 ± 4.5
C_f (0.5*) + SF	0.48%	$(1.5 \pm 0.1) \times 10^4$	$+1.45 \pm 0.09$	-0.89 ± 0.09	$+1.45 \pm 0.09$	-0.89 ± 0.09
C_f (1.0*) + SF	0.95%	$(8.3 \pm 0.5) \times 10^2$	$+2.82 \pm 0.11$	$+0.48 \pm 0.11$	$+2.82 \pm 0.11$	$+0.48 \pm 0.11$
C_f (0.5*) + L	0.41%	$(9.7 \pm 0.6) \times 10^4$	$+1.20 \pm 0.05$	-1.14 ± 0.05	$+1.20 \pm 0.05$	-1.14 ± 0.05
C_f (1.0*) + L	0.82%	$(1.8 \pm 0.2) \times 10^3$	$+2.10 \pm 0.08$	-0.24 ± 0.08	$+2.10 \pm 0.08$	-0.24 ± 0.08

* % by weight of cement
SF: silica fume
L: latex

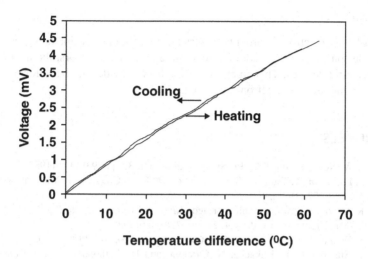

FIGURE 5.14 Variation of the Seebeck voltage (with copper as the reference) vs. the temperature difference during heating and cooling for steel-fiber silica-fume cement paste containing steel fibers in the amount of 1.0% by weight of cement.

Table 5.2 shows that the volume electrical resistivity is much higher for the steel-fiber cement pastes than the corresponding carbon-fiber cement pastes. This is attributed to the much lower volume fraction of fibers in the former. An increase in the steel or carbon fiber content from 0.5% to 1.0% by weight of cement decreases resistivity, though the decrease is more significant for the carbon fiber case than for the steel fiber case. That the resistivity decrease is not large when the steel fiber content is increased from 0.5% to 1.0% by weight of cement, and that the resistivity is still high at a steel fiber content of 1.0%, suggest that the steel fiber content of 1.0% is below the percolation threshold.

Whether with or without silica fume or latex, the change of the Seebeck voltage with temperature is more reversible and linear at a steel fiber content of 1.0% by weight of cement than at 0.5%. This is attributed to the larger role of the cement matrix at the lower steel fiber content, and the contribution of the cement matrix to irreversibility and nonlinearity. Irreversibility and nonlinearity are particularly significant when the cement paste contains no fiber.

From a practical point of view, the steel-fiber silica-fume cement paste containing steel fibers in the amount of 1.0% by weight of cement is particularly attractive for use in temperature sensing, as the magnitude of the absolute thermoelectric power is the highest (-68 μV/°C), and the variation of the Seebeck voltage with the temperature difference between the hot and cold ends is reversible and linear. The magnitude of the absolute thermoelectric power is as high as those of commercial thermocouple materials.

A cement-based thermocouple in the form of a junction between dissimilar cement pastes exhibits thermocouple sensitivity 70 ± 7 μV/°C.[68] The dissimilar cement pastes are steel fiber cement paste (n-type) and carbon-fiber silica-fume cement paste (p-type). The junction is made by pouring the cement pastes side by side.

5.6 CONCLUSION

Cement-based electronics provide electrical circuit elements, including conductor and diode (pn-junction), in addition to strain and damage sensors, thermistors, and thermoelectric devices. This emerging technology provides the basis for the use of concrete structures for electronic functions.

REFERENCES

1. J. W. Newman, Int. SAMPE Symp. Exhib., Vol. 32, pp. 938-944 (1987).
2. S. Furukawa, Y. Tsuji and S. Otani, Proc. 30th Japan Congress on Materials Research, pp. 149-152 (1987).
3. K. Saito, N. Kawamura and Y. Kogo, Advanced Materials: The Big Payoff, National SAMPE Technical Conf., Vol. 21, pp. 796-802 (1989).
4. S. Wen and D.D.L. Chung, Cem. Concr. Res., 29(3), 445-449 (1999).
5. T. Sugama, L. E. Kukacka, N. Carciello and D. Stathopoulos, Cem. Concr. Res., 19(3), 355-365 (1989).
6. X. Fu, W. Lu and D.D.L. Chung, Cem. Concr. Res., 26(7), 1007-1012 (1996).
7. X. Fu, W. Lu and D.D.L. Chung, Carbon, 36(9), 1337-1345 (1998).
8. Y. Xu and D.D.L. Chung, Cem. Concr. Res., 29(5), 773-776 (1999).
9. T. Yamada, K. Yamada, R. Hayashi and T. Herai, Int. SAMPE Symp. Exhib., Vol. 36, Pt. 1, pp. 362-371 (1991).
10. T. Sugama, L. E. Kukacka, N. Carciello and B. Galen, Cem. Concr. Res., 18(2), 290-300 (1988).
11. B. K. Larson, L. T. Drzal and P. Sorousian, Composites, 21(3), 205-215 (1990).
12. A. Katz, V.C. Li and A. Kazmer, J. Mater. Civil Eng., 7(2), 125-128 (1995).
13. S. B. Park and B. I. Lee, Cem. Concr. Composites, 15(3), 153-163 (1993).
14. P. Chen, X. Fu and D.D.L. Chung, ACI Mater. J., 94(2), 147-155 (1997).
15. P. Chen and D.D.L. Chung, Composites, 24(1), 33-52 (1993).
16. A. M. Brandt and L. Kucharska, Materials for the New Millennium, Proc. Mater. Eng. Conf., Vol. 1, pp. 271-280 (1996).
17. H. A. Toutanji, T. El-Korchi, R. N. Katz and G. L. Leatherman, Cem. Concr. Res., 23(3), 618-626 (1993).
18. N. Banthia and J. Sheng, Cem. Concr. Composites, 18(4), 251-269 (1996).
19. H. A. Toutanji, T. El-Korchi and R. N. Katz, Com. Concr. Composites, 16(1), 15-21 (1994).
20. S. Akihama, T. Suenaga and T. Banno, Int. J. Cem. Composites Lightweight Concr., 6(3), 159-168 (1984).
21. M. Kamakura, K. Shirakawa, K. Nakagawa, K. Ohta and S. Kashihara, Sumitomo Metals (1983).
22. A. Katz and A. Bentur, Cem. Concr. Res., 24(2), 214-220 (1994).
23. Y. Ohama and M. Amano, Proc. 27th Japan Congress on Materials Research, pp. 187-191 (1983).
24. Y. Ohama, M. Amano and M. Endo, Concr. Int.: Design Constr., 7(3), 58-62 (1985).
25. K. Zayat and Z. Bayasi, ACI Mater. J., 93(2), 178-181 (1996).
26. P. Soroushian, F. Aouadi and M. Nagi, ACI Mater. J., 88(1), 11-18 (1991).
27. B. Mobasher and C. Y. Li, ACI Mater. J., 93(3), 284-292 (1996).
28. N. Banthia, A. Moncef, K. Chokri and J. Sheng, Can. J. Civ. Eng., 21(6), 999-1011 (1994).

29. B. Mobasher and C. Y. Li, *Infrastructure: New Materials and Methods of Repair*, Proc. Mater. Eng. Conf., No. 804, pp. 551-558 (1994).
30. P. Soroushian, M. Nagi and J. Hsu, *ACI Mater. J.*, 89(3), 267-276 (1992).
31. P. Soroushian, *Constr. Specifier*, 43(12), 102-108 (1990).
32. A. K. Lal, *Batiment Int./Building Research & Practice*, 18(3), 153-161 (1990).
33. S. B. Park, B. I. Lee and Y. S. Lim, *Cem. Concr. Res.*, 21(4), 589-600 (1991).
34. S. B. Park and B. I. Lee, *High Temperatures — High Pressures*, 22(6), 663-670 (1990).
35. P. Soroushian, Mohamad Nagi and A. Okwuegbu, *ACI Mater. J.*, 89(5), 491-494 (1992).
36. M. Pigeon, M. Azzabi and R. Pleau, *Cem. Concr. Res.*, 26(8), 1163-1170 (1996).
37. N. Banthia, K. Chokri, Y. Ohama and S. Mindess, *Adv. Cem. Based Mater.*, 1(3), 131-141 (1994).
38. N. Banthia, C. Yan and K. Sakai, *Com. Concr. Composites*, 20(5), 393-404 (1998).
39. T. Urano, K. Murakami, Y. Mitsui and H. Sakai, *Composites — Part A: Applied Science & Manufacturing*, 27(3), 183-187 (1996).
40. A. Ali and R. Ambalavanan, *Indian Concr. J.*, 72(12), 669-675 (1998).
41. P. Chen, X. Fu and D.D.L. Chung, *Cem. Concr. Res.*, 25(3), 491-496 (1995).
42. M. Zhu and D.D.L. Chung, *Cem. Concr. Res.*, 27(12), 1829-1839 (1997).
43. M. Zhu, R. C. Wetherhold and D.D.L. Chung, *Cem. Concr. Res.*, 27(3), 437-451 (1997).
44. P. Chen and D.D.L. Chung, *Smart Mater. Struct.*, 2, 22-30 (1993)
45. P. Chen and D.D.L. Chung, *Composites*, Part B, 27B, 11-23 (1996).
46. P. Chen and D.D.L. Chung, *J. Am. Ceram. Soc.*, 78(3), 816-818 (1995).
47. D.D.L. Chung, *Smart Mater. Struct.*, 4, 59-61 (1995).
48. P. Chen and D.D.L. Chung, *ACI Mater. J.*, 93(4), 341-350 (1996).
49. X. Fu and D.D.L. Chung, *Cem. Concr. Res.*, 26(1), 15-20 (1996).
50. X. Fu, E. Ma, D.D.L. Chung and W. A. Anderson, *Cem. Concr. Res.*, 27(6), 845-852 (1997).
51. X. Fu and D.D.L. Chung, *Cem. Concr. Res.*, 27(9), 1313-1318 (1997).
52. X. Fu, W. Lu and D.D.L. Chung, *Cem. Concr. Res.*, 28(2), 183-187 (1998).
53. D.D.L. Chung, *TANSO*, (190), 300-312 (1999).
54. Z. Shi and D.D.L. Chung, *Cem. Concr. Res.*, 29(3), 435-439 (1999).
55. Q. Mao, B. Zhao, D. Sheng and Z. Li, *J. Wuhan U. Tech.*, Mater. Sci. Ed., 11(3), 41-45 (1996).
56. Q. Mao, B. Zhao, D. Shen and Z. Li, *Fuhe Cailiao Xuebao/Acta Materiae Compositae Sinica*, 13(4), 8-11 (1996).
57. M. Sun, Q. Mao and Z. Li, *J. Wuhan U. Tech.*, Mater. Sci. Ed., 13(4), 58-61 (1998).
58. B. Zhao, Z. Li and D. Wu, *J. Wuhan Univ. Tech.*, Mater. Sci. Ed., 10(4), 52-56 (1995).
59. S. Wen and D.D.L. Chung, *Cem. Concr. Res.*, 30(8), 1289-1294 (2000).
60. S. Wen and D.D.L. Chung, *Cem. Concr. Res.*, 31(2), 297-301 (2001).
61. S. Wen and D.D.L. Chung, *Cem. Concr. Res.*, 29(6), 961-965 (1999).
62. W.J. McCarter, *J. Am. Ceram. Soc.*, 78, 411-415 (1995).
63. M. Sun, Z. Li, Q. Mao and D. Shen, *Cem. Concr. Res.* 28(4), 549-554 (1998).
64. M. Sun, Z. Li, Q. Mao and D. Shen, *Cem. Concr. Res.* 28(12), 1707-1712 (1998).
65. S. Wen and D.D.L. Chung, *Cem. Concr. Res.* 29(12), 1989-1993 (1999).
66. S. Wen and D.D.L. Chung, *Cem. Concr. Res.* 30(8), 1295-1298 (2000).
67. S. Wen and D.D.L. Chung, *J. Mater. Res.* 15(12), (2000).
68. S. Wen and D.D.L. Chung, *Cem. Concr. Res.*, 31(3), 507-510 (2001).
69. D. Bontea, D.D.L. Chung and G.C. Lee, *Cem. Concr. Res.* 30(4), 651-659 (2000).
70. S. Wen and D.D.L. Chung, *Cem. Concr. Res.*, in press.

71. J. Lee and G. Batson, Materials for the New Millennium, Proc. 4th Mater. Eng. Conf., Vol. 2, pp. 887-896 (1996).

72. X. Fu and D.D.L. Chung, *ACI Mater. J.,* 96(4), 455-461 (1999).

73. Y. Xu and D.D.L. Chung, *Cem. Concr. Res.,* 29(7), 1117-1121 (1999).

74. Y. Shinozaki, Adv. Mater.: Looking Ahead to the 21st Century, Proc. 22nd National SAMPE Tech. Conf., Vol. 22, pp. 986-997 (1990).

75. X. Fu and D.D.L. Chung, *Cem. Concr. Res.,* 25(4), 689-694 (1995).

76. J. Hou and D.D.L. Chung, *Cem. Concr. Res.,* 27(5), 649-656 (1997).

77. G. G. Clemena, *Materials Performance,* 27(3), 19-25 (1988).

78. R. J. Brousseau and G. B. Pye, *ACI Mater. J.,* 94(4), 306-310 (1997).

79. P. Chen and D.D.L. Chung, *Smart Mater. Struct.,* 2, 181-188 (1993).

80. P. Chen and D.D.L. Chung, *J. Electron. Mater.,* 24(1), 47-51 (1995).

81. X. Wang, Y. Wang and Z. Jin, *Fuhe Cailiao Xuebao/Acta Materiae Compositae Sinica,* 15(3), 75-80 (1998).

82. N. Banthia, S. Djeridane and M. Pigeon, *Cem. Concr. Res.,* 22(5), 804-814 (1992).

83. P. Xie, P. Gu and J. J. Beaudoin, *J. Mater. Sci.,* 31(15), 4093-4097 (1996).

84. Z. Shui, J. Li, F. Huang and D. Yang, *J. Wuhan Univ. Tech.,* Mater. Sci. Ed., 10(4), 37-41 (1995).

85. S. Wen and D.D.L. Chung, *Cem. Concr. Res.,* 31(2), 141-147 (2001).

86. X. Fu and D.D.L. Chung, *Cem. Concr. Res.,* 28(6), 795-801 (1998).

87. X. Fu and D.D.L. Chung, *Carbon,* 36(4), 459-462 (1998).

88. X. Fu and D.D.L. Chung, *Cem. Concr. Res.,* 26(10), 1467-1472 (1996); 27(2), 314 (1997).

89. T. Fujiwara and H. Ujie, *Tohoku Kogyo Daigaku Kiyo, 1: Rikogakuhen,* No. 7, 179-188 (1987).

90. Y. Shimizu, A. Nishikata, N. Maruyama and A. Sugiyama, *Terebijon Gakkaishi/J. Inst. Television Engineers of Japan,* 40(8), 780-785 (1986).

91. P. Chen and D.D.L. Chung, *ACI Mater. J.,* 93(2), 129-133 (1996).

92. T. Uomoto and F. Katsuki, *Doboku Gakkai Rombun-Hokokushu/Proc. Japan Soc. Civil Engineers,* No. 490, Pt. 5-23, 167-174 (1994-1995).

93. C. M. Huang, D. Zhu, C. X. Dong, W. M. Kriven, R. Loh and J. Huang, *Ceramic Eng. Sci. Proc.,* 17(4), 258-265 (1996).

94. C. M. Huang, D. Zhu, X. Cong, W. M. Kriven, R. R. Loh and J. Huang, *J. Am. Ceramic Soc.,* 80(9), 2326-2332 (1997).

95. T-J. Kim and C-K. Park, *Cem. Concr. Res.,* 28(7), 955-960 (1998).

96. S. Igarashi and M. Kawamura, *Doboku Gakkai Rombun-Hokokushu/Proc. Japan Soc. Civil Eng.,* No. 502, Pt. 5-25, 83-92 (1994).

97. M. Z. Bayasi and J. Zeng, *ACI Structural J.,* 94(4), 442-446 (1997).

98. G. Campione, S. Mindess and G. Zingone, *ACI Mater. J.,* 96(1), 27-34 (1999).

99. T. Yamada, K. Yamada and K. Kubomura, *J. Composite Mater.,* 29(2), 179-194 (1995).

100. S. Delvasto, A. E. Naaman and J. L. Throne, *Int. J. Cem. Composites Lightweight Concr.,* 8(3), 181-190 (1986).

101. C. Park, Nippon Seramikkusu Kyokai, *Gakujutsu Ronbunshi — J. Ceramic Soc. Japan,* 106(1231), 268-271 (1998).

102. Y. Shao, S. Marikunte and S. P. Shah, *Concr. Int.,* 17(4), 48-52 (1995).

103. S. Wen and D.D.L. Chung, *Cem. Concr. Res.,* 30(4), 661-664 (2000).

104. S. Wen and D.D.L. Chung, *Cem. Concr. Res.,* 31(2), 129-133 (2001).

6 Self-Sensing of Carbon Fiber Polymer-Matrix Structural Composites

CONTENTS

SYNOPSIS The self-sensing of continuous carbon fiber polymer-matrix structural composites has been attained by electrical measurements. The sensing pertains to strain, damage, temperature, bond degradation, structural transitions, and the fabrication process. The experimental methods and data interpretation are covered in this chapter.

RELEVANT APPENDICES: *A, B, C, F, G, J*

6.1 INTRODUCTION

Polymer-matrix composites for structural applications typically contain continuous fibers such as carbon, polymer, and glass fibers; continuous fibers tend to be more effective than short fibers as a reinforcement. Polymer-matrix composites with continuous carbon fibers are used for aerospace, automobile, and civil structures.

Because carbon fibers are electrically conducting, whereas polymer and glass fibers are not, carbon fiber composites exhibit electrical properties that depend on parameters such as strain, damage, and temperature, thereby attaining the ability to sense themselves through electrical measurement. The self-sensing ability is valuable for smart structures, smart manufacturing, structural vibration control, and structural health monitoring. Self-sensing means the elimination of attached or embedded sensors, as the structural material itself is the sensor. The consequence is reduced cost, enhanced durability, larger sensing volume, and the absence of mechanical property degradation.

This chapter is focused on the use of DC electrical measurements for self-sensing in continuous carbon fiber polymer-matrix structural composites. The sensing is in terms of the strain, damage, temperature, bond degradation, structural transitions, and the composite fabrication process.

6.2 BACKGROUND

Carbon fibers are electrically conducting, while the polymer matrix is electrically insulating (except for the rare situation in which the polymer is an electrically conducting one[1]). The continuous fibers in a composite laminate are in the form of layers called laminae. Each lamina comprises many bundles (called tows) of fibers in a polymer matrix. Each tow consists of thousands of fibers. There may or may not be twists in a tow. Each fiber has a diameter ranging from 7 to 12 μm. The tows within a lamina are typically oriented in the same direction, but tows in different laminae may or may not be in the same direction. A laminate with tows in all the laminae oriented in the same direction is said to unidirectional. A laminate with tows in adjacent laminae oriented at an angle of 90° is said to be cross-ply. In general, an angle of 45° and other angles may be involved for various laminae. This is desired for attaining the mechanical properties required for the laminate in various directions in the plane of the laminate.

Within a lamina with tows in the same direction, electrical conductivity is highest in the fiber direction. In the transverse direction in the plane of the lamina, the conductivity is not zero, even though the polymer matrix is insulating. This is because there are contacts between fibers of adjacent tows.[2] In other words, a fraction of the fibers of one tow touches a fraction of the fibers of an adjacent tow here and there along the length of the fibers. These contacts result from the fact that fibers are not perfectly straight or parallel (even though the lamina is said to be unidirectional), and that the flow of the polymer matrix (or resin) during composite fabrication can cause a fiber to be not completely covered by the polymer or resin (even though, prior to composite fabrication, each fiber may be completely covered by the polymer or resin, as in the case of a prepreg, i.e., a fiber sheet impregnated with the polymer or resin). Fiber waviness is known as marcelling. Transverse conductivity gives information on the number of fiber-fiber contacts in the plane of the lamina.

For similar reasons, the contacts between fibers of adjacent laminae cause the conductivity in the through-thickness direction (direction perpendicular to the plane of the laminate) to be nonzero. Thus, the through-thickness conductivity gives information on the number of fiber-fiber contacts between adjacent laminae.

Matrix cracking between the tows of a lamina decreases the number of fiber-fiber contacts in the plane of the lamina, thus decreasing transverse conductivity. Similarly, matrix cracking between adjacent laminae (as in delamination[3]) decreases the number of fiber-fiber contacts between adjacent laminae, thus decreasing through-thickness conductivity. This means that the transverse and through-thickness conductivities can indicate damage in the form of matrix cracking.

Fiber damage (as distinct from fiber fracture) decreases the conductivity of a fiber, thereby decreasing longitudinal conductivity. However, due to the brittleness of carbon fibers, the decrease in conductivity because of fiber damage prior to fiber fracture is rather small.[4]

Fiber fracture causes a much larger decrease in the longitudinal conductivity of a lamina than does fiber damage. If there is only one fiber, a broken fiber results in an open circuit, i.e., zero conductivity. However, a lamina has a large number of fibers, and adjacent fibers can make contact here and there. Therefore, the portions of a broken fiber still contribute to the longitudinal conductivity of the lamina. As a result, the decrease in conductivity due to fiber fracture is less than what it would be if a broken fiber did not contribute to the conductivity. Nevertheless, the effect of fiber fracture on longitudinal conductivity is significant, so longitudinal conductivity can indicate damage in the form of fiber fracture.[5]

The through-thickness volume resistance of a laminate is the sum of the volume resistance of each of the laminae in the through-thickness direction, and the contact resistance of each of the interfaces between adjacent laminae (i.e., the interlaminar interface). For example, a laminate with eight laminae has eight volume resistances and seven contact resistances, all in the through-thickness direction. Thus, to study the interlaminar interface, it is better to measure the contact resistance between two laminae rather than the through-thickness volume resistance of the entire laminate.

Measurement of the contact resistance between laminae can be made by allowing two laminae (strips) to contact at a junction, and using the two ends of each strip for making four electrical contacts.[6] An end of the top strip and an end of the bottom strip serve as contacts for passing current. The other end of the top strip and the other end of the bottom strip serve as contacts for voltage measurement. The fibers in the two strips can be in the same direction or in different directions. This method is a form of the four-probe method of electrical resistance measurement. The configuration is illustrated in Figure 6.1 for cross-ply and unidirectional laminates. To make sure that the volume resistance within a lamina in the through-thickness direction does not contribute to the measured resistance, the fibers at each end of a lamina strip should be electrically shorted together by using silver paint or other conducting media. The measured resistance is the contact resistance of the junction. This resistance, multiplied by the area of the junction, gives the contact resistivity, which is independent of the area of the junction and just depends on the nature of the interlaminar interface. The unit of the contact resistivity is $\Omega.m^2$, whereas that of the volume resistivity is $\Omega.m$.

The structure of the interlaminar interface is more prone to change than the structure within a lamina. For example, damage in the form of delamination is much more common than damage in the form of fiber fracture. Moreover, the structure of the interlaminar interface is affected by the interlaminar stress, which is particularly

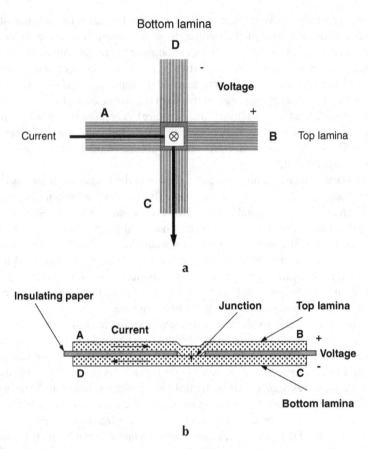

FIGURE 6.1 Specimen configuration for measurement of the contact electrical resistivity between laminae: (**a**) cross-ply laminae, (**b**) unidirectional laminae.

significant when the laminae are not unidirectional (as the anisotropy within each lamina enhances interlaminar stress). The structure of the interlaminar interface also depends on the extent of consolidation of the laminae during composite fabrication. The contact resistance provides a sensitive probe of the structure of the interlaminar interface.

The measurement of the volume resistivity in the through-thickness direction can be conducted by using the four-probe method, in which each of the two current contacts is in the form of a conductor loop on each of the two outer surfaces of the laminate in the plane of the laminate, and each of the two voltage contacts is in the form of a conductor dot within the loop.[3] An alternate method is to have four of the laminae in the laminate be extra long to serve as electrical leads. The two outer leads are for current contacts; the two inner leads are for voltage contacts. The use of a thin metal wire inserted at an end into the interlaminar space during composite fabrication to serve as an electrical contact is not recommended because the quality of the electrical contact between the metal wire and carbon fibers is hard to control, and the wire is intrusive to the composite. The alternate method is less convenient than the method involving loops and dots, but it approaches more closely the ideal four-probe method.

6.3　SENSING STRAIN

Smart structures that can monitor their own strain are valuable for structural vibration control. Self-monitoring of strain has been achieved in carbon fiber epoxy-matrix composites without the use of an embedded or attached sensor,[7-11] as the electrical resistance of the composite in the through-thickness or longitudinal direction changes reversibly with longitudinal strain due to change in the degree of fiber alignment. Tension in the fiber direction increases the degree of fiber alignment, thereby decreasing the chance for fibers of adjacent laminae to touch one another. As a consequence, the through-thickness resistance increases while the longitudinal resistance decreases.

Figure 6.2[9] shows the change in longitudinal resistance during cyclic longitudinal tension in the elastic regime for a unidirectional continuous carbon fiber epoxy-matrix composite with eight fiber layers. The stress amplitude is equal to 14% of the breaking stress. The strain returns to zero at the end of each cycle. Because of the small strains involved, the fractional resistance change $\Delta R/R_0$ is essentially equal to the fractional change in resistivity. The longitudinal $\Delta R/R_0$ decreases upon loading and increases upon unloading in every cycle, such that R irreversibly decreases slightly after the first cycle (i.e., $\Delta R/R_0$ does not return to zero at the end of the first cycle). At higher stress amplitudes the effect is similar, except that both the reversible and irreversible parts of $\Delta R/R_0$ are larger.

Figure 6.3[9] shows the change in the through-thickness resistance during cyclic longitudinal tension in the elastic regime for the same composite. The stress amplitude is equal to 14% of the breaking stress. The through-thickness $\Delta R/R_0$ increases upon loading and decreases upon unloading in every cycle, such that R irreversibly decreases slightly after the first cycle (i.e., $\Delta R/R_0$ does not return to zero at the end of the first cycle). Upon increasing the stress amplitude, the effect is similar, except that the reversible part of $\Delta R/R_0$ is larger.

The strain sensitivity (gage factor) is defined as the reversible part of $\Delta R/R_0$ divided by the longitudinal strain amplitude. It is negative (from -18 to -12) for the longitudinal $\Delta R/R_0$, and positive (17–24) for the through-thickness $\Delta R/R_0$. The magnitudes are comparable for the longitudinal and through-thickness strain sensitivities. As a result, whether the longitudinal R or the through-thickness R is preferred for strain sensing depends on the convenience of electrical contact application for the geometry of the particular smart structure.

Figure 6.4[9] shows the compressive stress, strain, and longitudinal $\Delta R/R_0$ obtained simultaneously during cyclic compression at stress amplitudes equal to 14% of the breaking stress for a similar composite having 24 rather than 8 fiber layers. The longitudinal $\Delta R/R_0$ increases upon compressive loading and decreases upon unloading in every cycle, such that resistance R irreversibly increases very slightly after the first cycle. The magnitude of the gage factor is lower in compression (-1.2) than in tension (from -18 to -12).

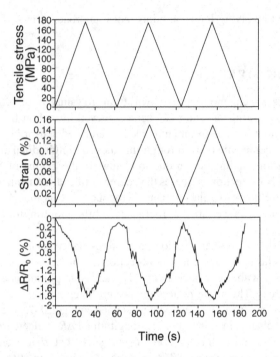

FIGURE 6.2 Longitudinal stress and strain and fractional resistance increase ($\Delta R/R_0$) obtained simultaneously during cyclic tension at a stress amplitude equal to 14% of the breaking stress for continuous fiber epoxy-matrix composite.

A dimensional change without any resistivity change would have caused longitudinal R to increase during tensile loading and decrease during compressive loading. In contrast, the longitudinal R decreases upon tensile loading and increases upon compressive loading. In particular, the magnitude of $\Delta R/R_0$ under tension is 7 to 11 times that of $\Delta R/R_0$ calculated by assuming that $\Delta R/R_0$ is due only to dimensional change and not to any resistivity change. Hence the contribution of $\Delta R/R_0$ from dimensional change is negligible compared to that from resistivity change.

The irreversible behavior, though small compared to the reversible behavior, is such that R (longitudinal or through-thickness) under tension is irreversibly decreased after the first cycle. This behavior is attributed to the irreversible disturbance to the fiber arrangement at the end of the first cycle such that the fiber arrangement becomes less neat. A less neat fiber arrangement means a greater chance for the adjacent fiber layers to touch one another.

6.4 SENSING DAMAGE

Self-monitoring of damage has been achieved in continuous carbon fiber polymer-matrix composites, as the electrical resistance of the composite changes with damage.[3,8,12-25] Minor damage in the form of slight matrix damage and/or disturbance to the fiber arrangement is indicated by the longitudinal and through-thickness resis-

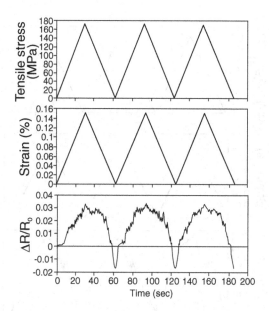

FIGURE 6.3 Longitudinal stress and strain and the through-thickness $\Delta R/R_0$ obtained simultaneously during cyclic tension at a stress amplitude equal to 14% of the breaking stress for continuous fiber epoxy-matrix composite.

tance decreasing irreversibly because of increase in the number of contacts between fibers, as shown after one loading cycle in Figures 6.2 to 6.4. More significant damage in the form of delamination or interlaminar interface degradation is indicated by the through-thickness resistance (or, more exactly, the contact resistivity of the interlaminar interface) increasing due to decrease in the number of contacts between fibers of different laminae. Major damage in the form of fiber breakage is indicated by the longitudinal resistance increasing irreversibly.

During mechanical fatigue, delamination was observed to begin at 30% of the fatigue life, whereas fiber breakage was observed to begin at 50% of the fatigue life. Figure 6.5[12] shows an irreversible resistance increase occurring at about 50% of the fatigue life during tension-tension fatigue testing of a unidirectional continuous carbon fiber epoxy-matrix composite. The resistance and stress are in the fiber direction. The reversible changes in resistance are due to strain, which causes the resistance to decrease reversibly in each cycle, as in Figure 6.2.

Figure 6.6[13] shows the variation of the contact resistivity with temperature during thermal cycling. The temperature is repeatedly increased to various levels. A group of cycles in which the temperature amplitude increases cycle by cycle and then decreases cycle by cycle back to the initial low temperature amplitude is referred to as a group. Figure 6.6(a) shows the results of the first ten groups, while Figure 6.6(b) shows the first group only. The contact resistivity decreases upon heating in every cycle of every group. At the highest temperature (150°C) of a group, a spike of resistivity increase occurs, as shown in Figure 6.6(b). It is attributed to damage at the interlaminar interface. In addition, the baseline resistivity (the top envelope)

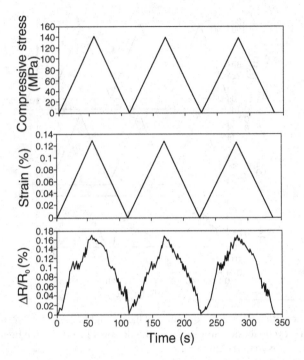

FIGURE 6.4 Longitudinal stress, strain, and $\Delta R/R_0$ obtained simultaneously during cyclic compression (longitudinal) at a stress amplitude equal to 14% of the breaking stress for continuous fiber epoxy-matrix composite.

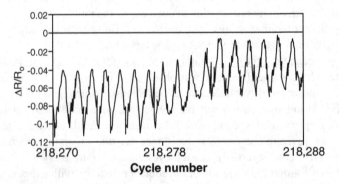

FIGURE 6.5 Variation of longitudinal $\Delta R/R_0$ with cycle number during tension-tension fatigue testing for a carbon fiber epoxy-matrix composite. Each cycle of reversible decrease in resistance is due to strain. The irreversible increase in resistance at around Cycle 218,281 is due to damage in the form of fiber breakage.

gradually and irreversibly shifts downward as cycling progresses, as shown in Figure 6.6(a). The baseline decrease is probably due to matrix damage within a lamina and the resulting decreases in modulus and residual stress.

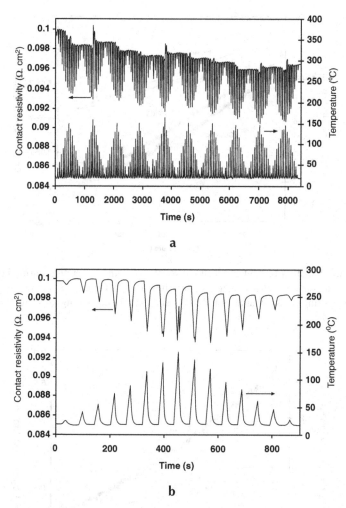

FIGURE 6.6 Variation of contact electrical resistivity with time and of the temperature with time during thermal cycling of a carbon fiber epoxy-matrix composite; (**b**) is the magnified view of the first 900 s of (**a**).

6.5 SENSING TEMPERATURE

Continuous carbon fiber epoxy-matrix composites provide temperature sensing by serving as thermistors[26,27] and thermocouples.[28]

The thermistor function stems from the reversible decrease on temperature of the contact electrical resistivity at the interface between fiber layers. Figure 6.7 shows the variation of the contact resistivity ρ_c with temperature during reheating and subsequent cooling, both at 0.15°C/min, for carbon fiber epoxy-matrix composites cured at 0 and 0.33 MPa. The corresponding Arrhenius plots of log contact conductivity (inverse of contact resistivity) versus inverse absolute temperature during heating are shown in Figure 6.8. From the slope of the Arrhenius plot, which is

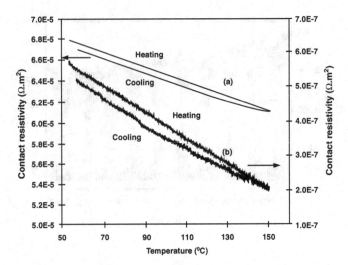

FIGURE 6.7 Variation of contact electrical resistivity with temperature during heating and cooling of carbon fiber epoxy-matrix composites at 0.15°C/min; (**a**) for composite made without any curing pressure, and (**b**) for composite made with a curing pressure of 0.33 MPa.

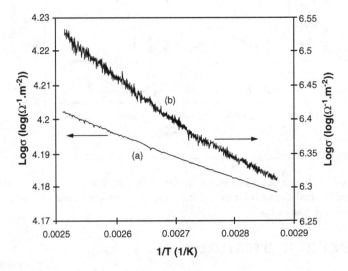

FIGURE 6.8 Arrhenius plot of log contact conductivity vs. inverse absolute temperature during heating of carbon fiber epoxy-matrix composites at 0.15°C/min (**a**) for composite made without any curing pressure, and (**b**) for composite made with curing pressure of 0.33 MPa.

quite linear, the activation energy can be calculated. The linearity of the Arrhenius plot means that the activation energy does not change throughout the temperature variation. This activation energy is the energy for electron jumping from one lamina to another. Electronic excitation across this energy enables conduction in the through-thickness direction.

TABLE 6.1

Activation Energy for Various Carbon Fiber Epoxy-Matrix Composites
(The standard deviations are shown in parentheses.)

Composite Configuration	Curing Pressure (MPa)	Composite Thickness (mm)	Contact Resistivity ρ_{co} ($\Omega.m^2$)	Activation Energy (kJ/mol)		
				Heating at 0.15°C/min	Heating at 1°C/min	Cooling at 0.15°C/min
Cross-ply	0	0.36	7.3×10^{-5}	1.26	1.24	1.21
				(2×10^{-3})	(3×10^{-3})	(8×10^{-4})
	0.062	0.32	1.4×10^{-5}	1.26	1.23	1.23
				(4×10^{-3})	(7×10^{-3})	(4×10^{-3})
	0.13	0.31	1.8×10^{-5}	1.62	1.57	1.55
				(3×10^{-3})	(4×10^{-3})	(2×10^{-3})
	0.19	0.29	5.4×10^{-6}	2.14	2.15	2.13
				(3×10^{-3})	(3×10^{-3})	(1×10^{-3})
	0.33	0.26	4.0×10^{-7}	11.4	12.4	11.3
				(4×10^{-2})	(8×10^{-2})	(3×10^{-2})
Unidirectional	0.42	0.23	2.9×10^{-5}	1.02	0.82	0.78
				(3×10^{-3})	(4×10^{-3})	(2×10^{-3})

The activation energies, thicknesses, and room temperature contact resistivities for samples made at different curing pressures and composite configurations are shown in Table 6.1. All the activation energies were calculated based on the data at 75 to 125°C. In this temperature regime, the temperature change was very linear and well controlled. From Table 6.1 it can be seen that, for the same composite configuration (cross-ply), the higher the curing pressure, the smaller the composite thickness (because of more epoxy being squeezed out), the lower the contact resistivity, and the higher the activation energy. A smaller composite thickness corresponds to a higher fiber volume fraction in the composite.

During curing and subsequent cooling, the matrix shrinks, so a longitudinal compressive stress will develop in the fibers. For carbon fibers, the modulus in the longitudinal direction is much higher than that in the transverse direction. Moreover, the carbon fibers are continuous in the longitudinal direction. Thus, the overall shrinkage in the longitudinal direction tends to be less than that in the transverse direction. Therefore, there will be a residual interlaminar stress in the two cross-ply layers in a given direction. This stress accentuates the barrier for the electrons to jump from one lamina to another.

After curing and subsequent cooling, heating will decrease thermal stress. Both thermal stress and curing stress contribute to residual interlaminar stress. Therefore, the higher the curing pressure, the larger the fiber volume fraction, the greater the residual interlaminar stress, and the higher the activation energy, as shown in Table 6.1. Besides the residual stress, thermal expansion can also affect contact resistance by changing the contact area. However, calculation shows that the contribution of thermal expansion is less than one-tenth of the observed change in contact resistance with temperature.

The electron jump occurs primarily at points where direct contact takes place between fibers of adjacent laminae. The direct contact is possible due to the flow of the epoxy resin during composite fabrication, and also to the slight waviness of the fibers, as explained in Reference 3, in relation to the through-thickness volume resistivity of a carbon fiber epoxy-matrix composite.

The curing pressure for the sample in the unidirectional composite configuration was higher than that of any of the cross-ply samples (Table 6.1). Consequently, the thickness was the lowest. As a result, the fiber volume fraction was the highest. However, the contact resistivity of the unidirectional sample was the second highest rather than the lowest, and its activation energy was the lowest rather than the highest. The low activation energy is consistent with the fact that there was no CTE or curing shrinkage mismatch between the two unidirectional laminae and, as a result, no interlaminar stress between them. This low value supports the notion that interlaminar stress is important in affecting activation energy. The high contact resistivity for the unidirectional case can be explained in the following way. In the cross-ply samples, the pressure during curing forced the fibers of the two laminae to press onto one another and contact tightly. In the unidirectional sample, the fibers of one of the laminae just sank into the other lamina at the junction, so pressure helped relatively little in the contact between fibers of adjacent laminae. Moreover, in the cross-ply situation, every fiber at the lamina-lamina interface contacted many fibers of the other lamina, while, in the unidirectional situation, every fiber had little chance to contact the fibers of the other lamina. Therefore, the number of contact points between the two laminae was less for the unidirectional sample than for the cross-ply samples.

The thermocouple function stems from the use of n-type and p-type carbon fibers (obtained by intercalation) in different laminae. The thermocouple sensitivity and linearity are as good as or better than those of commercial thermocouples. By using two laminae that are cross-ply, a two-dimensional array of thermistors or thermocouple junctions is obtained, thus allowing temperature distribution sensing.

Table 6.2 shows the Seebeck coefficient and the absolute thermoelectric power of carbon fibers, and the thermocouple sensitivity of epoxy-matrix composite junctions. A positive value of the absolute thermoelectric power indicates p-type behavior; a negative value indicates n-type behavior. Pristine P-25 is slightly n-type; pristine T-300 is strongly n-type. A junction comprising pristine P-25 and pristine T-300 has a positive thermocouple sensitivity that is close to the difference of the Seebeck coefficients of T-300 and P-25, whether the junction is unidirectional or cross-ply. Pristine P-100 and pristine P-120 are both slightly n-type. Intercalation with sodium causes them to become strongly n-type. Intercalation with bromine causes P-100 and P-120 to become strongly p-type. A junction comprising bromine-intercalated P-100 and sodium-intercalated P-100 has a positive thermocouple sensitivity that is close to the sum of the magnitudes of the absolute thermoelectric powers of the bromine-intercalated P-100 and the sodium-intercalated P-100. Similarly, a junction comprising bromine-intercalated P-120 and sodium-intercalated P-120 has a positive thermocouple sensitivity that is close to the sum of the

TABLE 6.2
Seebeck Coefficient (µV/°C) and Absolute Thermoelectric Power (µV/°C) of Carbon Fibers and Thermocouple Sensitivity (µV/°C) of Epoxy-Matrix Composite Junctions (All junctions are unidirectional unless specified as cross-ply. The temperature range is 20–110°C.)

	Seebeck Coefficient with Copper as the Reference (µV/°C)	Absolute Thermoelectric Power (µV/°C)	Thermocouple Sensitivity (µV/°C)
P-25*	+ 0.8	– 1.5	
T-300*	– 5.0	– 7.3	
P-25* + T-300*			+ 5.5
P-25* + T-300* (cross-ply)			+ 5.4
P-100*	– 1.7	– 4.0	
P-120*	– 3.2	– 5.5	
P-100 (Na)	– 48	– 50	
P-100 (Br₂)	+ 43	+ 41	
P-100 (Br₂) + P-100 (Na)	– 42	– 44	+ 82
P-120 (Na)	+ 38	+ 36	
P-120 (Br₂)			+ 74
P-120 (Br₂) + P-120 (Na)			

* Pristine (i.e., not intercalated)

magnitudes of their absolute thermoelectric powers. Figure 6.9 shows the linear relationship of the measured voltage with the temperature difference between hot and cold points for the junction comprising bromine-intercalated P-100 and sodium-intercalated P-100.

A junction comprising n-type and p-type partners has a thermocouple sensitivity that is close to the sum of the magnitudes of the absolute thermoelectric powers of the two partners. This is because the electrons in the n-type partner, as well as the holes in the p-type partner, move away from the hot point toward the corresponding cold point. As a result, the overall effect on the voltage difference between the two cold ends is additive.

By using junctions comprising strongly n-type and strongly p-type partners, a thermocouple sensitivity as high as +82 µV/°C has been attained. Semiconductors are known to exhibit much higher values of the Seebeck coefficient than metals, but the need to have thermocouples in the form of long wires makes metals the main materials for thermocouples. Intercalated carbon fibers exhibit much higher values of the Seebeck coefficient than metals. Yet, unlike semiconductors, their fiber and fiber composite forms make them convenient for practical use as thermocouples.

The thermocouple sensitivity of the carbon fiber epoxy-matrix composite junctions is independent of the extent of curing, and is the same for unidirectional and cross-ply junctions. This is consistent with the fact that the thermocouple effect hinges on the difference in the bulk properties of the two partners and is not an interfacial phenomenon. This means that the interlaminar interfaces in a fibrous

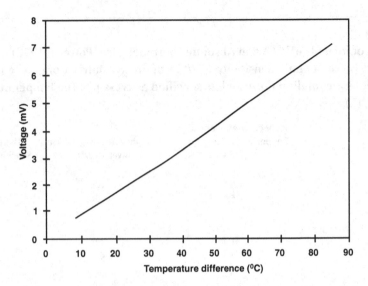

FIGURE 6.9 Variation of the measured voltage with the temperature difference between hot and cold points for the epoxy-matrix composite junction comprising bromine-intercalated P-100 and sodium-intercalated P-100 carbon fibers.

composite serve as thermocouple junctions in the same way, irrespective of the lay-up configuration of dissimilar fibers in the laminate. As a structural composite typically has fibers in multiple directions, this behavior facilitates the use of a structural composite as a thermocouple array.

It is important to note that thermocouple junctions do not require any bonding agent other than epoxy, which serves as the matrix of the composite and not as an electrical contact medium. In spite of the presence of the epoxy matrix in the junction area, direct contact occurs between a fraction of the fibers of a lamina and a fraction of the fibers of the other lamina, resulting in a conduction path in the direction perpendicular to the junction.

6.6 SENSING BOND DEGRADATION

Continuous fiber polymer-matrix composites are increasingly used to retrofit concrete structures, particularly columns.[28-30] The retrofit involves wrapping a fiber sheet around a concrete column, or placing a sheet on the surface of a concrete structure such that the sheet is adhered to the underlying concrete using a polymer, most commonly epoxy. This method is effective for the repair of even quite badly damaged concrete structures. Although the fibers and polymer are very expensive compared to concrete, the alternative of tearing down and rebuilding the concrete structure is often costlier. Both glass fibers and carbon fibers are used for composite retrofit. Glass fibers are advantageous for their relatively low cost, but carbon fibers are advantageous for their high tensile modulus.

The effectiveness of a composite retrofit depends on the quality of the bond between the composite and the underlying concrete, as good bonding is necessary

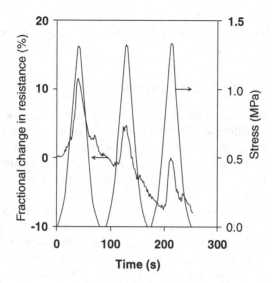

FIGURE 6.10 The fractional change in resistance for the fiber composite retrofit on a concrete substrate during cyclic compressive loading.

for load transfer. Peel testing for bond quality evaluation is destructive.[31] Nondestructive methods to evaluate the bond quality are valuable. They include acoustic methods, which are not sensitive to small amounts of debonding or bond degradation,[32] and dynamic mechanical testing.[33,34] Electrical resistance measurement can be used for nondestructive evaluation of the interface between concrete and its carbon fiber composite retrofit. The concept is that bond degradation causes the electrical contact between the carbon fiber composite retrofit and the underlying concrete to degrade. Since concrete is electrically more conducting than air, the presence of an air pocket at the interface causes an increase in the measured apparent volume resistance of the composite retrofit in a direction in the plane of the interface. Hence, bond degradation is accompanied by an increase in the apparent resistance of the composite retrofit. Although the polymer matrix is electrically insulating, the presence of a thin layer of epoxy at the interface is unable to electrically isolate the composite retrofit from the underlying concrete.

Figure 6.10 shows the fractional change in resistance during cyclic compressive loading at a stress amplitude of 1.3 MPa. The stress is along the fiber direction. Stress returns to zero at the end of each cycle. In each cycle, the electrical resistance increases reversibly during compressive loading. This is attributed to the reversible degradation of the bond between carbon fiber sheet and concrete substrate during compressive loading. This bond degradation decreases the chance for fibers to touch the concrete substrate, leading to a resistance increase.

As cycling progresses, both the maximum and minimum values of the fractional change in resistance in a cycle decrease. This is attributed to the irreversible disturbance in the fiber arrangement during repeated loading and unloading. This disturbance increases the chance for fibers to touch the concrete substrate, causing resistance to decrease irreversibly as cycling progresses.

As shown in Figure 6.10, the first cycle exhibits the highest value of the fractional change in resistance. This is because the greatest extent of bond degradation takes place during the first cycle.

6.7 SENSING STRUCTURAL TRANSITIONS

The polymer matrix of a composite material can undergo structural transitions such as glass transition, melting, cold-crystallization, and solid-state curing. Although the polymer matrix is insulating, the effect of a structural transition on fiber morphology (e.g., fiber waviness) results in an increase in the electrical resistivity of the composite in the fiber direction, allowing the resistance change to indicate a structural transition of the matrix.[35,36]

The glass transition and melting behavior of a thermoplastic polymer depends on the degree of crystallinity, the crystalline perfection, and other factors.[37-42] Knowledge of this is valuable for the processing and use of the polymer. This behavior is most commonly studied by differential scanning calorimetry (DSC),[37-42] although the DSC technique is limited to small samples, and the associated equipment is expensive and not portable. As the degree of crystallinity and the crystalline perfection of a polymer depend on the prior processing of the polymer and the effect of a process on the microstructure depends on the size and geometry of the polymer specimen, it is desirable to test the actual piece (instead of a small sample) for glass transition and melting behavior. Electrical resistance measurement provides a technique for this purpose.

DSC is a thermal analysis technique for recording the heat necessary to establish a zero temperature difference between a substance and a reference material, which are subjected to identical temperature programs in an environment heated or cooled at a controlled rate.[43] The recorded heat flow gives a measure of the amount of energy absorbed or evolved in a particular physical or chemical transformation, such as the glass transition, melting, or crystallization. The concept behind the electrical resistance technique is totally different from that of DSC. This technique involves measuring the DC electrical resistance when the polymer has been reinforced with electrically conducting fibers such as continuous carbon fibers. The resistance is in the fiber direction. The polymer molecular movements that occur at the glass transition and melting disturb the carbon fibers, which are much more conducting than the polymer matrix, and affect the electrical resistance of the composite in the fiber direction, allowing the resistance change to indicate glass transition and melting behavior.

Exposure of polyamides to heat and oxygen may cause changes in physical and chemical characteristics due to thermal oxidative degradation,[44] and thus changes in the mechanical properties. Prolonged annealing at a high temperature results in undesirable changes in the degree of crystallization and in the end groups, and may cause inter- and intramolecular transamidation reactions, chain scission, and cross-linking.[45-50] The electrical resistance technique is capable of studying the effect of annealing on the glass transition and melting behavior[35,36] in addition to studying the thermal history.[51]

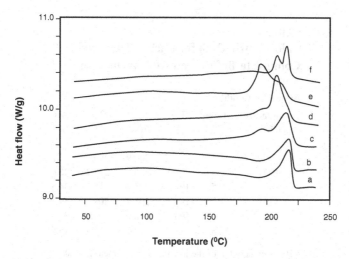

FIGURE 6.11 DSC thermograms showing the melting endothermic peaks before and after annealing at the temperatures and for the times shown. (**a**) As-received; (**b**) 100°C, 5 h; (**c**) 180°C, 5 h; (**d**) 180°C, 15 h; (**e**) 180°C, 30 h; (**f**) 200°C, 5 h.

6.7.1 DSC ANALYSIS

In Figure 6.11, (a) shows the DSC thermogram of the as-received carbon fiber Nylon-6-matrix composite.[35] The glass transition was not observed by DSC. T_m (melting temperature, as indicated by the peak temperature) is 218.5°C; (b), (c), (d), (e), and (f) show the effect of annealing time and temperature on the melting peak. The DSC results are summarized in Table 6.3.[34] Since T_m and ΔH of as-received and 100°C (5 h) annealed samples are almost the same (Figures 6.11(a) and (b)), it is attributed to the little change of the crystal perfection or the degree of crystallinity during annealing at 100°C for 5 h. Figure 6.11(c) shows the DSC thermogram of the sample annealed at 180°C for 5 h. It reveals two endothermic melting peaks with peak temperatures of 216 and 195°C. The lower temperature peak may be because of the structural reorganization during annealing in which the amorphous portion partly developed crystallinity.[35,52,53] As the annealing time increases to 15 h (Figure 6.11(d)), the high-temperature peak shifts to a lower temperature, but ΔH increases. As the annealing time increases to 30 h (Figure 6.11(e)), the height of the low-temperature peak increases, while that of the high-temperature peak decreases. These effects are probably due to the reorganization and thermal oxidative degradation of the Nylon-6 matrix. When the annealing time increases from 5 h (Figure 6.11(c)) to 15 h (Figure 6.11(d)), the degree of the crystallinity increases, so ΔH increases. However, at the same time, the extent of degradation increases due to thermal oxidation, which occurrs during annealing at a high temperature (180°C), resulting in lower crystal perfection. Therefore, the high-temperature peak shifts to a lower temperature. When the annealing time is long enough (30 h, Figure 6.11(e)), the crystalline portion from the reorganization process becomes dominant, as indi-

TABLE 6.3
Calorimetric Data for Carbon Fiber Nylon-6
Composite Before and After Annealing

| Annealing Condition | | | | | |
Temperature (°C)	Time (h)	T_{ml}^a (°C)	T_{onset}^b (°C)	T_m^c (°C)	ΔH^d (J/g)
e	e		200.9	218.5	26.7
100	5		205.5	218.2	26.6
180	5	194.8	201.3	215.5	34.8
180	15	196.3	201.4	208.9	39.1
180	30	196.3	200.0	209.0	38.6
200	5	208.3	212.9	216.4	16.5

[a] Peak temperature of the low-temperature melting peak
[b] Onset temperature of the high-temperature melting peak
[c] Peak temperature of the high-temperature melting peak
[d] Heat of fusion
[e] As-received

cated by the increase of the height of the low-temperature peak. When the sample has been annealed at 200°C for 5 h (Figure 6.11(f)), both T_m and ΔH decrease relative to the as-received sample. One possible explanation is that, when the annealing temperature is very high, the extent of thermal degradation is extensive, resulting in less crystalline perfection and a lower degree of crystallinity.

6.7.2 DC ELECTRICAL RESISTANCE ANALYSIS

Figure 6.12(a)[35] shows the fractional change in resistance for the as-received carbon fiber Nylon-6-matrix composite during heating, in which the temperature is raised from 25 to 350°C at a rate of 0.5°C/min. Two peaks were observed. The onset temperature of the first peak is 80°C, and that of the second peak is 220°C. The first peak is attributed to matrix molecular movement above T_g; the second peak is attributed to matrix molecular movement above T_m. Because the molecular movement above T_g is less drastic than that above T_m, the first peak is much lower than the second. As indicated before, the DSC thermogram of the as-received composite does not show a clear glass transition (Figure 6.11(a)). Therefore, the resistance is more sensitive to the glass transition than DSC. The onset temperature (220°C) of the second peak (Figure 6.12(a)) is higher than the onset temperature ($T_{onset} = 200.9$°C) of the DSC melting peak (Figure 6.11(a)), and is close to the melting temperature ($T_m = 218.5$°C) indicated by DSC (Figure 6.11(a)). The matrix molecular movement at T_{onset} is less intense than that at T_m, giving no effect on the resistance curve at T_{onset}. Another reason may be a time lag between the matrix molecular movement and the resistance change.

Figures 6.12(b), 6.12(c), and 6.12(d) show the effect of the annealing temperature. Comparison of Figures 6.12(a) and 6.12(b) shows that annealing at 100°C for

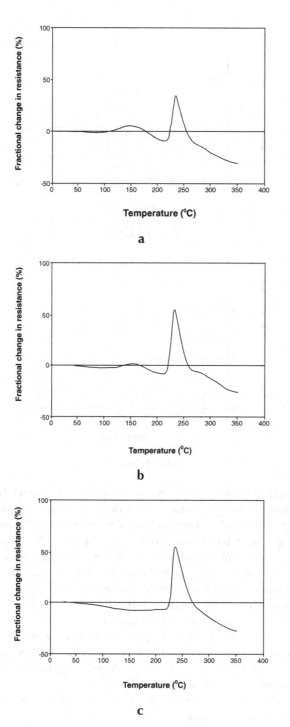

FIGURE 6.12 Effect of annealing condition on the variation of the electrical resistance with temperature. (a) As-received; (b) 100°C, 5 h; (c) 180°C, 5 h; (d) 200°C, 5 h; (e) 180°C, 15 h; (f) 180°C, 30 h.

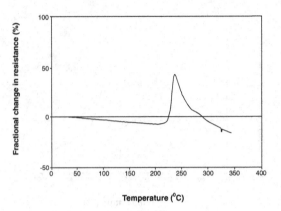

FIGURE 6.12d

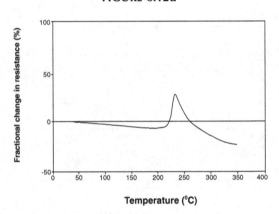

FIGURE 6.12e

5 h (Figure 6.12(b)) has little effect on the glass transition and melting behavior of the Nylon-6 matrix; this is consistent with the DSC results (Figures 6.11(a) and 6.11(b)). When the annealing temperature increases to 180°C (Figure 6.12(c)), the peak due to molecular movement above T_g disappears. This is attributed to the increase of the degree of crystallinity due to annealing. Because the crystalline portion has constraint on molecule mobility, the higher the degree of crystallinity, the lower the possibility of molecular movement above T_g.

The degree of crystallinity and the extent of thermal degradation affect molecular mobility above T_g. Figure 6.12(d) shows the fractional change in resistance of the sample annealed at 200°C for 5 h. No peak due to molecular movement above T_g was observed. The degree of crystallinity is less than that of the as-received sample, as shown by ΔH in Table 6.3; however, the higher extent of thermal degradation results in less molecular movement above T_g.

Figures 6.12(c) and 6.12(e) show the effect of annealing time from 5 to 15 h at 180°C. The height of the peak due to molecular movement above T_m decreases as the annealing time increases. A longer annealing time results in more thermal degradation of the matrix, which retards the molecular movement above T_m. That

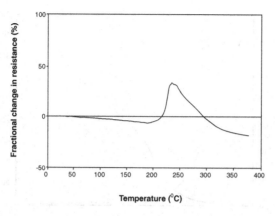

FIGURE 6.12f

this effect is due to a change of the extent of thermal degradation is also supported by the effect of annealing temperature, as shown in Figures 6.12(c) and 6.12(d). A higher annealing temperature likely enhances the extent of thermal degradation, resulting in a decrease of the height of the peak associated with molecular movement above T_m. Since the tail is more pronounced for samples with a larger extent of thermal degradation, as shown in Figures 6.12(d) and 6.12(f), it may be attributed to the lower molecular mobility caused by extensive thermal degradation.

6.8 SENSING COMPOSITE FABRICATION PROCESS

Continuous fiber polymer-matrix composites with thermosetting matrices are commonly made by stacking layers of fiber prepreg, and subsequent consolidation and curing under heat and pressure. Consolidation involves the use of pressure to bring the fiber layers closer to one another, and the use of heat to melt the resin in the prepreg so that the resin flow will allow the layers to come even closer together. A fraction of the resin may be squeezed out during consolidation. Curing occurs subsequent to consolidation and involves the resin completing its polymerization reaction so that it sets. Curing requires sufficient time and temperature in addition to the recommended pressure. There has been much work on the curing process,[54-59] but little or no work on the consolidation process. As consolidation is an important step in composite fabrication, understanding the consolidation process and characterizing the effectiveness of consolidation are valuable.

During consolidation, the thickness of the prepreg stack decreases. However, thickness change does not provide information on the extent of interaction between the prepreg layers. In the case of the fibers being carbon fibers, which are electrically conducting, the interaction between the prepreg layers leads to contact between fibers of adjacent layers, causing the volume electrical resistivity in the through-thickness direction to decrease. Hence, the resistivity provides information on the extent of fiber-fiber contact. In Reference 60, this resistivity was measured during consolidation for the purpose of studying the consolidation process in detail. Measurement of resistivity requires measurement of resistance as well as thickness.

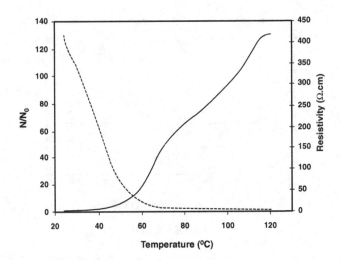

FIGURE 6.13 Variations of N/N_0 (solid line) and through-thickness electrical resistivity (dashed line) with temperature during consolidation heating to 120°C at a pressure of 0.56 MPa.

Figure 6.13[60] shows the variation of the through-thickness electrical resistivity and N/N_0 during consolidation at a pressure of 0.56 MPa. During consolidation, the temperature is raised linearly and reaches 120°C, the curing temperature. In Figure 6.13, the resistivity decreases and N/N_0 increases during consolidation such that the N/N_0 curve reveals three stages of consolidation. The first stage is characterized by a very gradual increase in N/N_0 (due to the solid form of the resin); the second stage is characterized by an abrupt increase in N/N_0 (due to the molten form of the resin); the third stage is characterized by a moderately gradual increase in N/N_0 (due to the thickening of the resin as the temperature increases). Similar changes in curvature of the N/N_0 plot were observed for consolidation conducted at ramped temperatures that reached 100, 110, 120, 130, or 140°C, although only the results for a maximum temperature of 120°C are shown in Figure 6.13. Similar effects were observed at a pressure of 1.10 MPa. The curve of N/N_0 vs. temperature is quantitatively quite independent of the maximum temperature of consolidation, but is quantitatively different for the two pressures. At the higher pressure, (1) N/N_0 reaches much higher values, (2) the second stage begins and ends at higher temperatures, and (3) the slope of the N/N_0 vs. temperature curve in the second stage is higher (7.18/°C for 1.10 MPa and 3.31/°C for 0.56 MPa). This means that pressure hastens consolidation and promotes the extent of consolidation, as expressed by the quantity N/N_0. In contrast, increasing the maximum temperature does not promote the extent of consolidation.

6.9 CONCLUSION

The measurement of the DC electrical resistance or voltage allows carbon fiber polymer-matrix structural composites to sense their own strain, damage, temperature, bond degradation, structural transitions, and fabrication process.

REFERENCES

1. X.B. Chen and D. Billaud, Ext. Abstr. Program — 20th Bienn. Conf. Carbon, pp. 274-275 (1991).
2. X. Wang and D.D.L. Chung, *Composite Interfaces*, 5(3), 191-199 (1998).
3. X. Wang and D.D.L. Chung, *Polym. Composites*, 18(6), 692-700 (1997).
4. X. Wang and D.D.L. Chung, *Carbon*, 35(5), 706-709 (1997).
5. X. Wang and D.D.L. Chung, *J. Mater. Res.*, 14(11), 4224-4229 (1999).
6. S. Wang and D.D.L. Chung, *Composite Interfaces*, 6(6), 497-506 (1999).
7. S. Wang and D.D.L. Chung *Polym. Composites*, 21(1), 13-19 (2000).
8. N. Muto, H. Yanagida, T. Nakatsuji, M. Sugita, Y. Ohtsuka and Y. Arai, *Smart Mater. Struct.*, 1, 324-329 (1992).
9. X. Wang, X. Fu and D.D.L. Chung, *J. Mater. Res.*, 14(3), 790-802 (1999).
10. X. Wang and D.D.L. Chung, *Composites: Part B*, 29B(1), 63-73 (1998).
11. P.E. Irving and C. Thiogarajan, *Smart Mater. Struct.*, 7, 456-466 (1998).
12. X. Wang, S. Wang and D.D.L. Chung, *J. Mater. Sci.*, 34(11), 2703-2714 (1999).
13. S. Wang and D.D.L. Chung, *Polym. Composites*, in press.
14. S. Wang and D.D.L. Chung, *Polym. Composites*, in press.
15. N. Muto, H. Yanagida, M. Miyayama, T. Nakatsuji, M. Sugita and Y. Ohtsuka, *J. Ceramic Soc. Japan*, 100(4), 585-588 (1992).
16. N. Muto, H. Yanagida, T. Nakatsuji, M. Sugita, Y. Ohtsuka, Y. Arai and C. Saito, *Adv. Composite Mater.*, 4(4), 297-308 (1995).
17. R. Prabhakaran, *Exp. Tech.*, 14(1), 16-20 (1990).
18. M. Sugita, H. Yanagida and N. Muto, *Smart Mater. Struct.*, 4(1A), A52-A57 (1995).
19. A.S. Kaddour, F.A.R. Al-Salehi, S.T.S. Al-Hassani and M.J. Hinton, *Composites Sci. Tech.*, 51, 377-385 (1994).
20. O. Ceysson, M. Salvia and L. Vincent, *Scripta Materialia*, 34(8), 1273-1280 (1996).
21. K. Schulte and Ch. Baron, *Composites Sci. Tech.*, 36, 63-76 (1989).
22. K. Schulte, *J. Physique IV*, C7(3), 1629-1636 (1993).
23. J.C. Abry, S. Bochard, A. Chateauminois, M. Salvia and G. Giraud, *Composites Sci. Tech.*, 59(6), 925-935 (1999).
24. A. Tedoroki, H. Kobayashi and K. Matuura, *JSME Int. J. Series A* — Solid Mechanics Strength of Materials, 38(4), 524-530 (1995).
25. S. Hayes, D. Brooks, T. Liu, S. Vickers and G.F. Fernando, *Proc. SPIE — Int. Soc. Optical Eng.*, Vol. 2718, pp. 376-384 (1996).
26. S. Wang and D.D.L. Chung, *Composite Interfaces*, 6(6), 497-506 (1999).
27. S. Wang and D.D.L. Chung, *Composites: Part B*, 30(6), 591-601 (1999).
28. S.Wang and D.D.L. Chung, *Composite Interfaces*, 6(6), 519-530 (1999).
29. H.A. Toutanji, *Composite Structures*, 44(2), 155-161 (1999).
30. H.A. Toutanji and T. El-Korchi, *J. Composites Constr.*, 3(1), 38-45 (1999).
31. Y. Lee and S. Matsui, Tech. Reports Osaka University, 48(2319-2337), 247-254 (1998).
32. V. M. Karbhari, M. Engineer and D. A. Eckel II, *J. Mater. Sci.*, 32(1), 147-156 (1997).
33. D.P. Henkel and J. D. Wood, *NDT & E International*, 24(5), 259-264 (1991).
34. A. K. Pandey and M. Biswas, *J. Sound Vib.*, 169(1), 3-17 (1994).
35. Z. Mei and D.D.L. Chung, *Polym. Composites*, 21(5), 711-715 (2000).
36. Z. Mei and D.D.L. Chung, *Polym. Composites*, 19(6), 709-713 (1998).
37. J.A. Kuphal, L.H. Sperling and L.M. Robeson, *J. Appl. Polym. Sci.*, 42, 1525-1535 (1991).
38. A.L. Simal and A.R. Martin, *J. Appl. Polym. Sci.*, 68, 453-474 (1998).

39. Ch.R. Davis, *J. Appl. Polym. Sci.*, 62, 2237-2245 (1996).
40. B.G. Risch and G.L. Wilkes, *Polymer*, 34, 2330-2343 (1993).
41. H.J. Oswald, E.A. Turi, P.J. Harget and Y.P. Khanna, *J. Macromol. Sci. Phys.*, B13(2), 231-254 (1977).
42. J.U. Otaigbe and W.G. Harland, *J. Appl. Polym. Sci.*, 36, 165-175 (1988).
43. M.E. Brown, *Introduction to Thermal Analysis: Techniques and Application*, Chapman & Hall, New York, p. 25 (1988).
44. C.H. Do, E.M. Pearce and B.J. Bulkin, *J. Polym. Sci.*, Part A: Polym. Chem., 25, 2409-2424 (1987).
45. M.C. Gupta and S.G. Viswanath, *J. Therm. Anal.*, 47(4), 1081-1091 (1996).
46. N. Avramova, *Polym. & Polym. Comp.*, 1(4), 261-274 (1993).
47. A.L. Simal and A.R. Martin, *J. Appl. Polym. Sci.*, 68, 441-452 (1998).
48. I.M. Fouda, M.M. El-Tonsy, F.M. Metawe, H.M. Hosny and K.H. Easawi, *Polym. Test.*, 17(7), 461-493 (1998).
49. I.M. Fouda, E.A. Seisa and K.A. El-Farahaty, *Polym. Test.*, 15(1), 3-12 (1996).
50. L.M. Yarisheva, L. Yu Kabal'nova, A.A. Pedy and A.L. Volynskii, *J. Therm. Anal.*, 38(5), 1293-1297 (1992).
51. Z. Mei and D.D.L. Chung, *Thermochim. Acta*, 369(1-2), 62-65 (2001).
52. Y.P. Khanna, *Macromolecules*, 25, 3298-3300 (1992).
53. Y.P. Khanna, *J. Appl. Polym. Sci.*, 40, 569-579 (1990).
54. C.W. Lee and B. Rice, *Int. SAMPE Symp. Exhib.*, 41(2), 1511-1517 (1996).
55. G.M. Maistros and I.K. Partridge, *Composites*, Part B, 29(3), 245-250 (1998).
56. R.P. Cocker, D.L. Chadwick, D.J. Dare and R.E. Challis, *Int. J. Adh. Adh.*, 18(5) 319-331 (1998).
57. G.R. Powell, P.A. Crosby, D.N. Waters, C.M. France, R.C. Spooncer and G.F. Fernando, *Smart Mater. Struct.*, 7(4), 557-568 (1998).
58. Y. Li and S. Menon, *Sensors*, 15(2), 14, 16, 18, 19 (1998).
59. R. Casalini, S. Corezzi, A. Livi, G. Levita and P.A. Rolla, *J. Appl. Polym. Sci.*, 65(1), 17-25 (1997).
60. S. Wang and D.D.L. Chung, *Polymer Composites*, 22(1), 42-46 (2001).

7 Structural Health Monitoring by Electrical Resistance Measurement

CONTENTS

SYNOPSIS Structural health monitoring by DC electrical resistance measurement is reviewed. The technique is valuable for evaluating composites and joints, provided the materials involved are not all electrically insulating. The measurement pertains to either the volume electrical resistivity of a bulk material or the contact resistivity of an interface, and it can be performed in real time during loading or heating.

RELEVANT APPENDICES: *A, B, C, G*

7.1 INTRODUCTION

Structural health monitoring refers to observing the integrity of a structure for the purpose of hazard mitigation, whether the hazard is due to live load, earthquake, wind, ocean waves, fatigue, aging, heat, or other factors. It mainly entails nondestructive sensing of the damage in the structure.

Extensive damage, such as large cracks on the surface, can be sensed by visual and liquid penetrant inspection. Both surface and subsurface defects can be sensed by magnetic particle inspection, eddy current testing, ultrasonic testing, x-radiography, and other methods. However, these techniques are not very sensitive to defects that are subtle in nature or microscopic in size. The resolution is particularly poor for magnetic particle inspection and x-radiography. The resolution is better for ultrasonic methods, but it is still limited to about 0.1 mm (depending on the frequency of the ultrasonic wave). For effective hazard mitigation, it is important to be able to detect defects when they are small.

Recent attention to structural health monitoring has been centered on the use of embedded or attached sensors, such as piezoelectric, optical fiber, microelectromechanical, acoustic, dynamic response, phase transformation, and other sensors,[1-38] and the use of tagging through the incorporation of piezoelectric, magnetic, or electrically conducting particles in the composite material.[39,40] The use of embedded sensors or particles suffers from the mechanical property degradation of the composite. Furthermore, embedded sensors are hard to repair. Attached sensors are easier to repair than embedded sensors, but they suffer from poor durability. Both embedded and attached sensors are much more expensive than the structural material, so they add a lot to the cost of the structure.

Electrical resistance measurement has received relatively little attention in terms of structural health monitoring. It does not involve any embedment or attachment, so it does not suffer from the problems described for embedded or attached sensors. Also, this method allows the entire structure to be monitored, whereas the use of embedded or attached sensors allows the structure to be monitored only at various places. The electrical resistance method is particularly effective for detecting small and subtle defects in composite materials and in joints. Materials such as carbon fiber polymer-matrix composites are lightweight structural composites that are important for aircraft, for which structural health monitoring is critical. Joints, whether by fastening or adhesion, are encountered in almost any structure, and their structural health monitoring is particularly challenging due to the difficulty of probing the joint interface.

7.2 CARBON FIBER POLYMER-MATRIX STRUCTURAL COMPOSITES

Polymer-matrix composites for structural applications typically contain continuous fibers such as carbon, polymer, and glass fibers because continuous fibers are more effective than short fibers as a reinforcement. Polymer-matrix composites with continuous carbon fibers (Section 6.2) are used for aerospace, automobiles, and civil structures.

The monitoring of damage has been achieved with continuous carbon fiber polymer-matrix composites, as the electrical resistance of the composite changes with damage (Section 6.4). In other words, the composites are sensors of their own damage.

7.3 CEMENT-MATRIX COMPOSITES

Concrete (a cement-matrix composite) is the dominant structural material for civil infrastructure. Particularly in the presence of short carbon fibers as an admixture, its electrical resistivity changes upon strain or damage, making it a sensor of its own strain (Section 5.4.1) and damage (Section 5.4.2). That both strain and damage can be sensed simultaneously through resistance measurement means that the strain/stress condition under which damage occurs can be obtained, thus facilitating damage origin identification. Damage is indicated by a resistance increase, which is larger and less reversible when the stress amplitude is higher. The resistance increase can be a sudden increase during loading or a gradual shift of baseline resistance.

7.4 JOINTS

Joining is one of the key processes in manufacturing and repair. It can be achieved by welding, diffusion bonding (autohesion in the case of polymers), soldering, brazing, adhesion, fastening, or other methods.

Joints can be evaluated destructively by mechanical testing, which involves debonding. However, it is preferred to use nondestructive methods, such as modulus (dynamic mechanical), acoustic, and electrical measurements. Electrical measurements are particularly attractive because of short response time and equipment simplicity. A requirement for the feasibility of electrical measurements for joint evaluation is that the components being joined are not electrical insulators. Thus, joints involving metals, cement (concrete), and conductor-filled polymers are suitable.

The method of electrical resistance measurement for joint evaluation most commonly involves measurement of the contact electrical resistivity of the joint interface. Contact resistivity is given by the product of the contact resistance and the joint area; it is a quantity that is independent of the joint area. Degradation of the joint causes contact resistivity to increase. A less common method involves measuring the apparent volume resistance of a component while component A is joined to component B. When B is less conducting than A, but is not insulating, degradation of the joint causes the apparent volume resistance of A to increase.

7.4.1 JOINTS INVOLVING COMPOSITE AND CONCRETE BY ADHESION

Continuous fiber polymer-matrix composites are used increasingly to retrofit concrete structures, particularly columns.[41-43] Retrofit involves wrapping a fiber sheet around a concrete column or placing a sheet on the surface of a concrete structure such that the fiber sheet is adhered to the underlying concrete by using a polymer, usually epoxy. This method is effective for the repair of even quite badly damaged concrete structures. Although the fibers and polymer are very expensive compared to concrete, the alternative of tearing down and rebuilding the structure is often more expensive than the composite retrofit. Both glass fibers and carbon fibers are used for composite retrofit. Glass fibers are advantageous for their relatively low cost, but carbon fibers are advantageous for their high tensile modulus.

The effectiveness of a composite retrofit depends on the quality of the bond between the composite and the underlying concrete, as good bonding is necessary for load transfer. Peel testing for bond quality evaluation is destructive.[44] Nondestructive methods to evaluate the bond quality are valuable. They include acoustic methods, which are not sensitive to small amounts of debonding or bond degradation,[45] and dynamic mechanical testing.[46] Electrical resistance measurement has been used for nondestructive evaluation of the interface between concrete and its carbon fiber composite retrofit.[41] This method is effective for studying the effects of temperature and debonding stress on the interface. The concept is that bond degradation causes the electrical contact between the carbon fiber composite retrofit and the underlying concrete to degrade. Since concrete is electrically more conducting than air, the presence of an air pocket at the interface causes the measured apparent volume resistance of the composite retrofit to increase in a direction in the plane of the interface. Hence, bond degradation is accompanied by an increase in the apparent resistance of the composite retrofit. Although the polymer matrix is electrically insulating, the presence of a thin layer of epoxy at the interface is unable to electrically isolate the composite retrofit from the underlying concrete.

The apparent resistance of the retrofit in the fiber direction is increased by bond degradation, whether the degradation is due to heat or stress. The degradation is reversible. Irreversible disturbance in the fiber arrangement occurs slightly as thermal or load cycling occurs, as indicated by the resistance decreasing cycle by cycle.[41]

7.4.2 JOINTS INVOLVING COMPOSITES BY ADHESION

Joining methods for polymers and polymer-matrix composites include autohesion, which is relevant to the self-healing of polymers. Diffusion bonding (or autohesion) involves interdiffusion among the adjoining materials in the solid state. In contrast, fusion bonding involves melting. Due to the relatively low temperatures of diffusion bonding compared to fusion bonding, diffusion bonding does not suffer from the undesirable side-effects that occur in fusion bonding, such as degradation and crosslinking of the polymer matrix. Although the diffusion bonding of metals has been widely studied, relatively little study has been conducted on the autohesion of polymers. Because of the increased segment mobility above the glass transition temperature (T_g), thermoplastics are able to undergo interdiffusion above T_g.

Diffusion, as a thermally activated process, takes time. How long diffusion takes depends on the temperature. In order for diffusion bonding or autohesion to be conducted properly, the kinetics of the process need to be known.

The study of the kinetics requires monitoring the process as it occurs. A real-time monitoring technique is obviously preferable to a traditional method that requires periodic interruption and cooling of the specimen. However, real-time monitoring is experimentally difficult compared with interrupted monitoring. The method described here is ideal for thermoplastic prepregs containing continuous carbon fibers, since the carbon fibers are conductive. Two carbon-fiber thermoplastic prepreg plies are placed together to form a joint. The electrical contact resistance of this joint is measured during autohesion. As autohesion occurs, the fibers in the plies being joined come closer together, resulting in a decrease in contact resistance. With the measurement of this resistance in real time, the autohesion process has been monitored as a function of time at different selected bonding temperatures for Nylon-6 and polyphenylenesulfide (PPS), both thermoplastics.[47] Arrhenius plots of a characteristic resistance decrease versus temperature allow determination of the activation energy for the process. This method can be used for monitoring the bonding of unfilled thermoplastics if a few carbon fibers are strategically placed.

Engineering thermoplastics can be bonded by autohesion above the glass transition temperature but below the melting temperature, or fusion welding (melting and solidification). Both methods involve heating and subsequent cooling. During cooling, the thermoplastic goes from a soft solid state (in the case of autohesion) or a liquid state (in the case of fusion welding) to a stiff state. If the thermoplastic members to be joined are anisotropic (as in the case of each member being reinforced with fibers) and the fiber orientation in the two members is not the same, the thermal expansion (actually contraction) mismatch at the bonding plane will cause thermal stress to build up during cooling. This thermal stress is detrimental to the quality of the adhesive bond.

Two scenarios can lead to the absence of bonding after cooling. One is the absence of bond formation at the high temperature during welding because of insufficient time or temperature. The other scenario is the presence of bonding at the high temperature, but the occurrence of debonding during cooling due to thermal stress. The cause of the absence of bonding is different in the two scenarios. In any given situation, the cause of the debonded joint must be understood if the absence of bonding after cooling is to be avoided.

The propensity for mutual diffusion in thermoplastic polymers increases with temperature. The contact at the interface across which interdiffusion takes place also plays a role. An intimate interface, as obtained by application of pressure to compress the two members together, also facilitates diffusion. Thus, the quality of the joint improves with increasing temperature and increasing pressure in the high-temperature period of welding. The poorer the quality of the joint attained at a high temperature, the greater the likelihood that thermal stress built up during subsequent cooling will be sufficient to cause debonding. Hence, merely having bonding achieved at the high temperature in welding is not enough. The bond must be of sufficient quality to withstand the abuse of thermal stress during cooling.

The quality of a joint is conventionally tested destructively by mechanical methods, or nondestructively by ultrasonic methods.[48,49] This testing is performed at room temperature after the joint has been cooled from the high temperature used in welding. As a result, the testing does not allow distinction between the two scenarios described above. The use of a nondestructive method, namely contact electrical resistance measurement, to monitor joint quality in real time during the high temperature period of welding, and also during subsequent cooling, has been shown.[50] The resistance increases by up to 600% upon debonding. The resistance increase is much greater than the resistance decrease during prior bonding. Debonding occurs during cooling when the pressure or temperature during prior bonding is not sufficiently high.

Adhesive joint formation between thermoplastic adherends typically involves heating to temperatures above the melting temperature (T_m) of the thermoplastic. During heating to the desired elevated temperature, time is spent in the range between the glass transition temperature (T_g) and the T_m. The dependence of the bond quality on the heating rate, heating time, and pressure has been investigated through measurement of the contact resistance between adherends in the form of carbon fiber-reinforced PPS.[51] A long heating time below the melting temperature (T_m) is detrimental to subsequent PPS adhesive joint development above T_m. This is due to curing reactions below T_m and consequent reduced mass flow response above T_m. A high heating rate enhances the bonding more than a high pressure.

7.4.3 Joints Involving Steels by Fastening

Mechanical fastening involves the application of a force to the components to be joined so as to prevent them from separating in service. During repair, maintenance, or other operations, unfastening may be needed. Hence, repeated fastening and unfastening may be necessary. By design, the stresses encountered by the components and fasteners are below the corresponding yield stresses so that no plastic deformation occurs. However, the local stress at the asperities at the interface can exceed the yield stress, resulting in local plastic deformation as shown for carbon steel fastened joints at a compressive stress of just 7% or less of the yield stress.[52] Plastic deformation results in changes in the joint interfaces. This means that the joint interface depends on the extent of prior fastening and unfastening. The joint interface affects the mechanical and corrosion behavior of the joint; this problem is of practical importance.

Stainless steel differs from carbon steel in the presence of a passive film, which is important to its corrosion resistance. The effect of repeated fastening and unfastening on the passive film is of concern.

Figures 7.1 (a) and (b)[53] show the variation in resistance and displacement during cyclic compressive loading of stainless steel on stainless steel at a stress amplitude of 14 MPa. In every cycle, the resistance decreases as the compressive stress increases such that the maximum stress corresponds to the minimum resistance and the minimum stress corresponds to the maximum resistance (Figure 7.1(a)).

The maximum resistance (in the unloaded condition) of every cycle increases upon stress cycling such that the increase is not significant until after 13 cycles (Figure 7.1(a)). The increase is due to the damage of the passive film and the consequent surface oxidation. The minimum resistance (at the maximum stress) of every cycle increases slightly upon cycling (Figure 7.1(b)), probably due to strain hardening.

The higher the stress amplitude, the fewer the number of stress cycles for passive film damage to begin. At the lowest stress amplitude of 3.5 MPa, passive film damage was not observed up to 30 cycles.

Comparison of the results on stainless steel and on carbon steel shows that the carbon steel joint is dominated by effects associated with plastic deformation, whereas the stainless steel joint is dominated by effects associated with passive film damage. The effect of the passive film is absent in the carbon steel joint, as expected from the absence of a passive film on carbon steel. The effects of plastic deformation

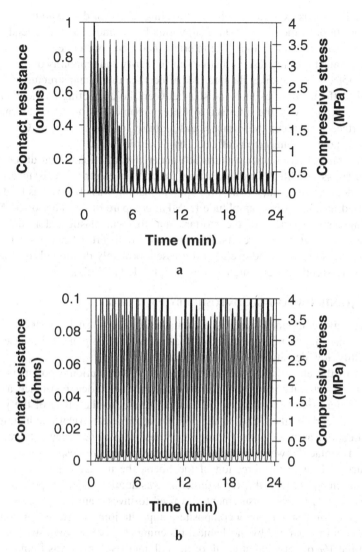

FIGURE 7.1 Variation of contact resistance (thick curve) and stress (thin curve) during cyclic compression of stainless steel on stainless steel at a stress amplitude of 14 MPa.

and strain hardening at asperities are much larger for carbon steel than for stainless steel, as expected from the lower yield stress of carbon steel.

7.4.4 JOINTS INVOLVING CONCRETE BY PRESSURE APPLICATION

Many concrete structures involve the direct contact of one cured concrete element with another such that one element exerts static pressure on the other due to gravity. In addition, dynamic pressure may be exerted by live loads on the structure. An example of such a structure is a bridge involving slabs supported by columns, with dynamic live loads exerted by vehicles traveling on the bridge. Another example is a concrete floor in the form of slabs supported by columns, with live loads exerted by people walking on the floor. The interface between concrete elements that are in pressure contact is of interest, as it affects the integrity and reliability of the assembly. For example, deformation at the interface affects the interfacial structure, which can affect the effectiveness of load transfer between the contacting elements, and can affect the durability of the interface to the environment. Moreover, deformation at the interface can affect the dimensional stability of the assembly. Of particular concern is how the interface is affected by dynamic loads.

A mortar-mortar contact was studied under dynamic loading at different compressive stress amplitudes by measuring the contact electrical resistance.[54,55] Irreversible decrease in contact resistance upon unloading was observed as load cycling progressed at a low stress amplitude (5 MPa, compared to a value of 64 MPa for the compressive strength of the mortar), due to local plastic deformation at the asperities at the interface. Irreversible increase in contact resistance at the maximum stress was observed as load cycling progressed, probably due to debris generation; it was more significant at a higher stress amplitude (15 MPa).

7.4.5 JOINTS INVOLVING COMPOSITES BY FASTENING

Fasteners and components are usually made of metals, such as steel. However, polymers are increasingly used for both fasteners and components because of their moldability, low density, and corrosion resistance.

Due to the electrically insulating behavior of conventional polymers and the need for an electrical conductor for measuring contact electrical resistance, a polymer with continuous carbon fibers in a direction parallel to the plane of the joint was used.[56] The carbon fibers caused the composite to be electrical conducting in the fiber direction, as well as the through-thickness direction, because there is some degree of contact between adjacent fibers in the composite despite the presence of the matrix.[57] Due to the direction of the fibers, the mechanical properties of the composite in the through-thickness direction was dominated by the polymer matrix.

Contact resistance measurement was used to investigate the effect of repeated fastening and unfastening on a composite-composite joint interface.[56] A composite-composite joint obtained by mechanical fastening at a compressive stress of 5% (or less) of the 1% offset yield strength of the polymer (Nylon-6) was found to exhibit irreversible decrease in contact electrical resistance upon repeated fastening and unfastening. The decrease occurred after up to 10 cycles of fastening and unfastening, although the decrease diminished with cycling. It was primarily due to local plastic deformation of the matrix at the asperities at the interface. Moreover, the stress required for the resistance to reach its minimum in a cycle decreased with cycling because of softening of the matrix.

7.5 CONCLUSION

DC electrical resistance measurement is useful for structural health monitoring, particularly in relation to composites and joints, provided the materials involved are not all electrically insulating. Monitoring involves measuring the volume resistivity of a bulk material or the contact resistivity of an interface. Measurement can be conducted in real time during loading or heating.

REFERENCES

1. V.K. Varadan and V.V. Varadan, *Proc. SPIE*, 3673, 359-368 (1999).
2. J.S. Vipperman, *AIAA/ASME/ASCE/AHS Structures*, Structural Dynamics and Materials Conference — Collection of Technical Papers, 4, 3107-3114 (1999).
3. R.C. Foedinger, D.L. Rea, J.S. Sirkis, C.S. Baldwin, J.R. Troll, R. Grande, C.S. Davis and T.L. VanDiver, *Proc. SPIE*, 3670, 289-301 (1999).
4. X. Deng, Q. Wang and V. Giurgiutiu, *Proc. SPIE*, 3668(I), 363-370 (1999).
5. Z. Jiang, K. Kabeya and S. Chonan, *Proc. SPIE*, 3668(I), 343-350 (1999).
6. C. Boller, C. Biemans, W.J. Staszewski, K. Worden and G.R. Tomlinson, *Proc. SPIE*, 3668(I), 285-294 (1999).
7. D.M. Castillo, C. Pardo de Vera and J.A. Guemes, *Key Eng. Mater.*, 167, 91-101 (1999).
8. Q. Wang and X. Deng, *Int. J. Solids Struct.*, 36(23), 3443-3468 (1999).
9. R. Pascual, J.C. Golinval and M. Razeto, Proceedings of the 17th International Modal Analysis Conference — IMAC, Vol. 1, pp. 238-243 (1999).
10. B.J. Maclean, M.G. Mladejovsky, M.R. Whitaker, M. Olivier and S.C. Jacobsen, Nondestructive Characterization of Materials in Aging Systems, Materials Research Society Symposium Proceedings, Vol. 503, pp. 309-320 (1998).
11. R. Foedinger, D. Rea, J. Sirkis, R. Wagreich, J. Troll, R. Grande, C. Davis and T.L. Vandiver, Proc. 43rd International SAMPE Symposium and Exhibition, Vol. 43, No. 1, pp. 444-457 (1998).
12. C.J. Groves-Kirkby, *GEC J. Tech.*, 15(1), 16-26 (1998).
13. S. Beard and F.-K. Chang, *J. Intelligent Mater. Syst. Struct.*, 8(10), 891-897 (1997).
14. G.W. Reich and K.C. Park, Collection of Technical Papers — AIAA/ASME/ASCE/ AHS/ASC Structures, Structural Dynamics and Materials Conference, Vol. 2, pp. 1653-1660 (1998).
15. D. Polla, L. Francis, W. Robbins and R. Harjani, Emerging Technologies for Machinery Health Monitoring and Prognosis, American Society of Mechanical Engineers, Tribology Division, TRIB., Vol. 7, pp. 19-24 (1997).
16. M.J. Schulz, A.S. Naser, S.K. Thyagarajan, T. Mickens and P.F. Pai, Proc. 16th International Modal Analysis Conference — IMAC, Vol. 1, pp. 760-766 (1998).
17. G.H. James, III, D.C. Zimmerman and R.L. Mayes, Proceedings of the 16th International Modal Analysis Conference — IMAC, Vol. 1, pp. 151-157 (1998).
18. D. Inaudi, N. Casanova, P. Kronenberg, S. Marazzi and S. Vurpillot, *Proc. SPIE*, 3044, 236-243 (1997).
19. P.J. Ellerbrock, *Proc. SPIE*, 3044, 207-218 (1997).
20. A.E. Aktan, A.J. Helmicki, V.J. Hunt, N. Catbas, M. Lenett and A. Levi, Proceedings of the American Control Conference, Vol. 2, pp. 873-877 (1997).
21. J.A.J. Fells, M.J. Goodwin, C.J. Groves-Kirkby, D.C.J. Reid, J.E. Rule, and M.B. Shell, *IEEE Colloquium* (Digest), (033), 10/1-10/11 (1997).
22. R.S. Ballinger and D.W. Herrin, 11th Biennial Conference on Reliability, Stress Analysis, and Failure Prevention, American Society of Mechanical Engineers, Design Engineering Division DE, Vol. 83, No. 2 (Pt. 1), pp. 181-187 (1995).
23. D.C. Zimmerman, S.W. Smith, H.M. Kim and T.J. Bartkowicz, *J. Vibr. Acoust. Trans. ASME*, 118(4), 543-550 (1996).

24. J.N. Schoess, G. Seifert, C.A. Paul, *Proc. SPIE*, 2718, 175-184 (1996).
25. J.N. Schoess, *Proc. SPIE*, 2717, 212-218 (1996).
26. J. Schoess, F. Malver, B. Iyer and J. Kooyman, 52nd Annual Forum Proceedings — American Helicopter Society, Vol. 2, pp. 1788-1793 (1996).
27. K.E. Castanien and C. Liang, *Proc. SPIE*, 2721, 38-49 (1996).
28. G. James, D. Roach, B. Hansche, R. Meza and N. Robinson, *Proc. Int. Conf. Eng., Constr., Oper. Space*, Vol. 2, pp. 1127-1133 (1996).
29. B.D. Westermo and L.D. Thompson, *Proc. SPIE*, 2443, 841-851 (1995).
30. D.C. Zimmerman, M. Kaouk and T. Simmermacher, *J. Mech. Design*, 117B, 214-221 (1995).
31. N. Narendran and J. Weiss, *Proc. Instrum. Soc. Am.*, Vol. 38, pp. 273-282 (1995).
32. D.A. Sofge, 2nd Australian and New Zealand Conference on Intelligent Information Systems — Proceedings, pp. 91-94 (1994).
33. C.A. Paul, G.P. Sendeckyj and G.P. Carman, *Proc. SPIE*, 1918, 154-164 (1993).
34. G. Park, H.H. Cudney and D.J. Inman, *Proc. SPIE*, 3670, 461-469 (1999).
35. G. Park, K. Kabeya, H.H. Cudney and D.J. Inman, *JSME Int. J. Series A — Solid Mech. Mater. Eng.*, 42(2), 249-258 (1999).
36. F. Lalande, C.A. Rogers, B.W. Childs, Z.A. Chaudhry, *Proc. SPIE*, 2717, 237-243 (1996).
37. B.D. Westermo and L.D. Thompson, *Proc. SPIE*, 2446, 37-46 (1995).
38. L. Thompson, B. Westermo and R. Waldbusser, Proceedings of the 42nd International Instrumentation Symposium, Instrument Society of America, pp. 567-576 (1996).
39. S.R. White, *Proc. SPIE*, 2442, 337-348 (1995).
40. R.F. Quattrone and J.B. Berman, Materials for the New Millennium, Proceedings of the Materials Engineering Conference, Vol. 2, pp. 1045-1054 (1996).
41. Z. Mei and D.D.L. Chung, *Cem. Concr. Res.*, 30(5), 799-802 (2000).
42. H.A. Toutanji, *Composite Struct.*, 44(2), 155-161 (1999).
43. H.A. Toutanji and T. El-Korchi, *J. Composites Constr.*, 3(1), 38-45 (1999).
44. V.M. Karbhari, M. Engineer and D.A. Eckel, II, *J. Mater. Sci.*, 32(1), 147-156 (1997).
45. D.P. Henkel, J.D. Wood, *NDT & E International*, 24(5), 259-264 (1991).
46. A.K. Pandey and M. Diswas, *J. Sound Vibr.*, 169(1), 3-17 (1994).
47. Z. Mei and D.D.L. Chung, *Int. J. Adh. Adh.*, 20, 173-175 (2000).
48. S. Dixon, C. Edwards and S.B. Palmer, *Ultrasonics*, 32(6), 425-430 (1994).
49. V.K. Winston, Proc. 43rd International SAMPE Symposium and Exhibition, Vol. 43, No. 2, pp. 1428-1437 (1998).
50. Z. Mei and D.D.L. Chung, *Int. J. Adh. Adh.*, 20, 135-139 (2000).
51. Z. Mei and D.D.L. Chung, *Int. J. Adh. Adh.*, 20, 273-277 (2000).
52. X. Luo and D.D.L. Chung, *J. Mater. Eng. Perf.*, 9, 95-97 (2000).
53. X. Luo and D.D.L. Chung, *Adv. Eng. Mater.*, 3(1-2), 62-65 (2001).
54. X. Luo and D.D.L. Chung, *Cem. Concr. Res.*, 30(2), 323-326 (2000).
55. X. Luo and D.D.L. Chung, *J. Mater. Sci.*, 35(19), 4795-4802 (2000).
56. X. Luo and D.D.L. Chung, *Polym. Eng. Sci.*, 40(7), 1505-1509 (2000).
57. X. Wang and D.D.L. Chung, *Polym. Composites*, 18(6), 692-700 (1997).

8 Modification of the Surface of Carbon Fibers for Use as a Reinforcement in Composite Materials

CONTENTS

SYNOPSIS Surface modification can greatly enhance the effectiveness of carbon fibers as a reinforcement in composite materials. The methods of surface modification, as used for polymer-matrix and metal-matrix composites, are covered.

RELEVANT APPENDIX: *J*

8.1 INTRODUCTION TO SURFACE MODIFICATION

The surface of a material greatly affects many of its properties. Modification of the surface is an effective way to improve material properties. Surface modification refers to modification of surface composition or surface structure. The surface region can have a thickness ranging from Å to μm. As modification of the surface is often simpler and less expensive than modification of the bulk, it is used technologically to improve materials for many applications.

Methods of surface modification include heating in a reactive environment (to change the surface composition or to form surface functional groups, a surface alloy, a surface compound, surface porosity, surface roughness, or a coating); heating the surface in an inert environment, such as by using a laser (to change the surface microstructure or crystallinity); bombardment by charged particles such as ions (plasma) or electrons (to introduce surface defects or roughness); immersion in a liquid, possibly followed by heating (to form a coating, or a reaction product in case the liquid is reactive); immersion in a solvent (to remove contaminants such as grease); subjecting the material to electrochemical oxidation (to form surface functional groups, surface roughness, etc.); and sandblasting (to roughen the surface).

Carbon (as in carbon fibers) can be modified by heating in air, as the oxygen in the air reacts with the surface carbon atoms, thereby forming surface functional groups in the form of C–O and/or C=O. Similar effects can be attained by subjecting carbon to electrochemical oxidation or to an oxygen plasma. Carbon can also be modified by heating in CO_2 gas, as the reaction

$$C + CO_2 \rightarrow 2CO$$

causes the conversion of solid carbon on the surface to CO gas, resulting in surface porosity — a form of carbon called activated carbon. Carbon can also be modified by immersion in silicon liquid to form SiC on the surface through the reaction

$$C + Si \rightarrow SiC$$

It can also be modified by heating in silane (SiH_4) gas to form SiC on the surface through the reaction

$$C + SiH_4 \rightarrow SiC + H_2$$

As carbon is typically either amorphous or partly crystalline, it can be modified by heating the surface in an inert atmosphere so that the surface crystallinity is increased — a process known as surface annealing. As sandblasting involves bombardment with abrasive particles, it cannot be used on surfaces of microscopic size (such as the surface of carbon fibers).

This chapter focuses on the surface modification of carbon fibers for use as a reinforcement in composite materials. The fibers are also used in the activated form for fluid purification by adsorption. In addition, carbon fibers are used as battery electrode materials and as materials for EMI shielding. Surface modification significantly improves carbon fibers for all these applications, although the dominant application is reinforcement.

8.2 INTRODUCTION TO CARBON FIBER COMPOSITES

Composite materials are materials containing more than one phase such that the different phases are artificially blended together. They are not multiphase materials in which the different phases are formed naturally by reactions, phase transformations, or other phenomena.

A composite material typically consists of one or more fillers in a certain matrix. A carbon fiber composite is one in which at least one of the fillers is carbon fibers, either short or continuous, unidirectional or multidirectional, woven or nonwoven. The matrix is usually a polymer, a metal, a carbon, a ceramic, or a combination of different materials. Except for sandwich composites, the matrix is three-dimensionally continuous, whereas the filler can be three-dimensionally discontinuous or continuous. Carbon fiber fillers are usually three-dimensionally discontinuous unless the fibers are three-dimensionally interconnected by weaving or by the use of a binder such as carbon.

The high strength and modulus of carbon fibers makes them useful as a reinforcement for polymers, metals, carbons, and ceramics, even though they are brittle. Effective reinforcement requires good bonding between the fibers and the matrix, especially for short fibers. For a unidirectional composite, the longitudinal tensile strength is quite independent of the fiber-matrix bonding, but the transverse tensile strength and the flexural strength increase with increasing fiber-matrix bonding. On the other hand, excessive fiber-matrix bonding can cause a composite with a brittle matrix (e.g., carbon and ceramics) to become more brittle. The strong fiber-matrix bonding causes cracks to propagate straightly in the direction perpendicular to the fiber-matrix interface without being deflected to propagate along this interface. In the case of a composite with a ductile matrix (metals and polymers), a crack initiating in the brittle fiber tends to be blunted when it reaches the ductile matrix, even when the fiber-matrix bonding is strong. Therefore, an optimum degree of fiber-matrix bonding is needed for brittle-matrix composites, whereas a high degree of fiber-matrix bonding is preferred for ductile-matrix composites.

The mechanisms of fiber-matrix bonding include chemical bonding, van der Waals bonding, and mechanical interlocking. Chemical bonding gives the largest bonding force, provided the density of chemical bonds across the fiber-matrix interface is sufficiently high. This density can be increased by chemical treatments of the fibers or by sizings on the fibers. Mechanical interlocking between the fibers and the matrix is an important contribution to the bonding if the fibers form a three-dimensional network. Otherwise, the fibers should have a rough surface in order for a small degree of mechanical interlocking to take place.

Both chemical and van der Waals bonding require the fibers to be in intimate contact with the matrix. For intimate contact to take place, the matrix or matrix precursor must be able to wet the surfaces of the carbon fibers during its infiltration into the carbon fiber preform. Chemical treatments and coatings can be applied to the fibers to enhance wetting. The choice of treatment or coating depends on the matrix. Another way to enhance wetting is to use high pressure during infiltration. A third method is to add a wetting agent to the matrix or matrix precursor before infiltration. As the wettability may vary with temperature, the infiltration temperature can be chosen to enhance wetting.

The occurrence of a reaction between the fibers and the matrix helps the wetting and bonding between the fibers and the matrix. An excessive reaction degrades the fibers, and the reaction product(s) may be undesirable for the mechanical, thermal, or moisture resistance properties of the composite. An optimum amount of reaction is preferred.

Carbon fibers are electrically and thermally conductive, in contrast to the non-conducting nature of polymer and ceramic matrices. Therefore, carbon fibers can serve not only as a reinforcement, but also as an additive for enhancing electrical or thermal conductivity. Furthermore, carbon fibers have nearly zero coefficient of thermal expansion, so they can also serve as an additive for lowering thermal expansion. The combination of high thermal conductivity and low thermal expansion makes carbon fiber composites useful for heat sinks in electronics, and for space structures that require dimensional stability. As the thermal conductivity of carbon fibers increases with the degree of graphitization, applications requiring a high thermal conductivity should use the graphitic fibers, such as high-modulus pitch-based fibers and vapor-grown carbon fibers.

Carbon fibers are more cathodic than practically any metal, so in a metal matrix, a galvanic couple is formed with the metal as the anode. This causes corrosion of the metal. The corrosion product tends to be unstable in moisture and causes pitting, which aggravates corrosion. To alleviate this problem, carbon fiber metal-matrix composites are often coated.

Carbon is the matrix that is most compatible with carbon fibers. The carbon fibers in a carbon-matrix composite (called carbon-carbon composite) serve to strengthen the composite, as the carbon fibers are much stronger than the carbon matrix because of the crystallographic texture in each fiber. Moreover, the carbon fibers serve to toughen the composite, as debonding between the fibers and the matrix provides a mechanism for energy absorption during mechanical deformation. In addition to having attractive mechanical properties, carbon-carbon composites are more thermally conductive than carbon fiber polymer-matrix composites. However, at elevated temperatures (above 320°C), carbon-carbon composites degrade due to the oxidation of carbon (especially the carbon matrix), which forms CO_2 gas. To alleviate this problem, carbon-carbon composites are coated.

Carbon fiber ceramic-matrix composites are more oxidation resistant than carbon-carbon composites. The most common form of such composites is carbon fiber-reinforced concrete. Although the oxidation of carbon is catalyzed by an alkaline environment and concrete is alkaline, the chemical stability of carbon fibers in concrete is superior to that of competitive fibers, such as polypropylene, glass, and steel. Composites containing carbon fibers in more advanced ceramic matrices (such as SiC) are rapidly being developed.

Carbon fiber composites are most commonly fabricated by the impregnation of the matrix or matrix precursor in the liquid state into the fiber preform, which is usually in the form of a woven fabric. In the case of composites in the shape of tubes, the fibers may be impregnated in the form of a continuous bundle from a spool and, subsequently, the bundles may by wound on a mandrel. Instead of impregnation, the fibers and matrix material may be intermixed in the solid state by commingling carbon fibers and matrix fibers, by coating the carbon fibers with the matrix material, by sandwiching carbon fibers with foils of the matrix material, or in other ways. After impregnation or intermixing, consolidation is carried out, often under heat and pressure.

Because of the decreasing price of carbon fibers, the applications of carbon fiber composites are rapidly widening to include the aerospace, automobile, marine, construction, biomedical, and other industries. This situation poses an unusual demand on research and development in the field of carbon fiber composites.

8.3 SURFACE MODIFICATION OF CARBON FIBERS FOR POLYMER-MATRIX COMPOSITES

Polymer-matrix composites are much easier to fabricate than metal-matrix, carbon-matrix, and ceramic-matrix composites, whether the polymer is a thermoset or a thermoplastic. This is because of the relatively low processing temperatures required for fabricating polymer-matrix composites. For thermosets, such as epoxy, phenolic, and furfuryl resin, the processing temperature ranges from room temperature to about 200°C; for thermoplastics, such as polyimide (PI), polyethersulfone (PES), polyetheretherketone (PEEK), polyetherimide (PEI), and polyphenyl sulfide (PPS), the processing temperature ranges from 300 to 400°C.

Thermosets (especially epoxy) have long been used as polymer matrices for carbon fiber composites. During curing, usually performed in the presence of heat and pressure, a thermoset resin hardens gradually due to the completion of polymerization and the cross-linking of the polymer molecules. Thermoplastics have recently become important because of their greater ductility and processing speed compared to thermosets, and the recent availability of thermoplastics that can withstand high temperatures. The higher processing speed of thermoplastics is due to the fact that thermoplastics soften immediately upon heating above the glass transition temperature (T_g), and the softened material can be shaped easily. Subsequent cooling completes the processing. In contrast, the curing of a thermoset resin is a reaction that occurs gradually.

Epoxy is by far the most widely used polymer matrix for carbon fibers. Trade names of epoxy include Epon®, Epirez®, and Araldite®. Epoxy has an excellent combination of mechanical properties and corrosion resistance, is dimensionally stable, exhibits good adhesion, and is relatively inexpensive. Moreover, the low molecular weight of uncured epoxide resins in the liquid state results in exceptionally high molecular mobility during processing. This mobility enables the resin to quickly wet the surface of carbon fiber, for example.

Epoxy resins are characterized by having two or more epoxide groups per molecule. The chemical structure of an epoxide group is:

$$CH_2 \overset{\displaystyle O}{\diagup \quad \diagdown} \underset{\underset{H}{|}}{C} {-}$$

The mers (repeating units) of typical thermoplasts used for carbon fibers are shown below, where Be = benzene ring.

PI

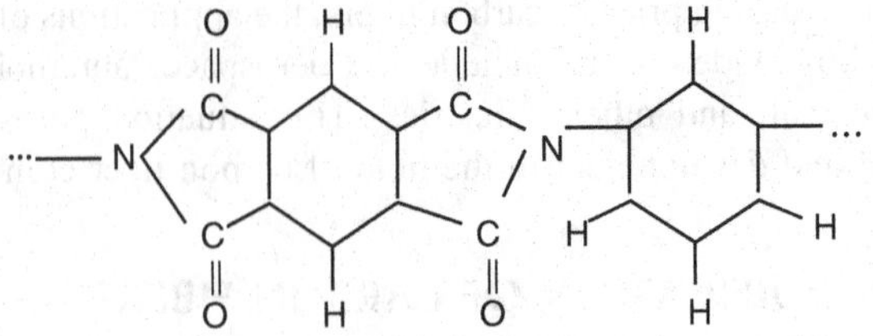

PEEK   ...—O – Be – O – Be – C – Be —...

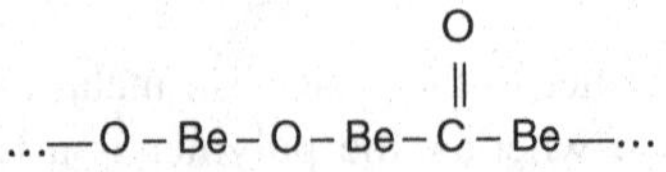

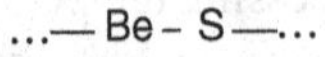

PPS   ...— Be – S —...

PES   ...— Be – O – Be – S —...

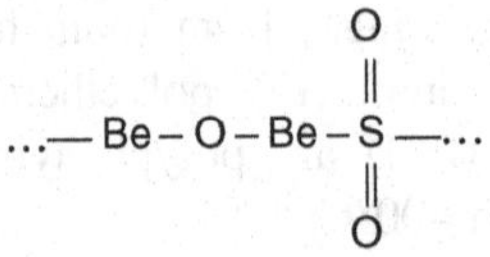

PEI

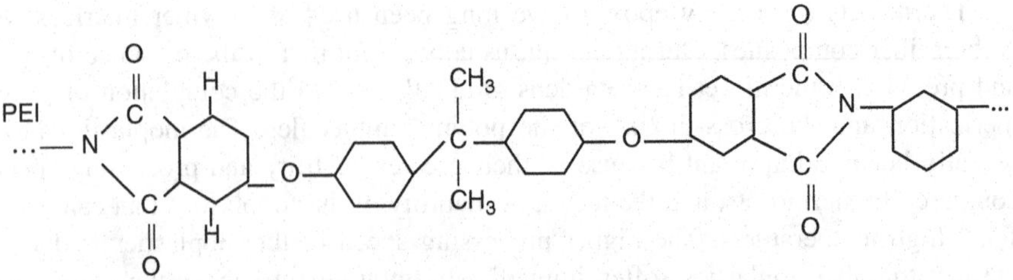

The properties of the above thermoplastics are listed in Table 8.1.[1-3] In contrast, epoxies have tensile strengths of 30–100 MPa, moduli of elasticity of 2.8–3.4 GPa, ductilities of 0–6%, and a density of 1.25 g/cm^3. Thus, epoxies are much more brittle than PES, PEEK, and PEI. In general, the ductility of a semicrystalline thermoplastic decreases with increasing crystallinity. For example, the ductility of PPS can range from 2 to 20%, depending on the crystallinity.[4] Another major difference between thermoplastics and epoxies lies in the higher processing temperatures of thermoplastics (300–400°C).

Surface treatments of carbon fibers are essential for improving the bonding between the fibers and the polymer matrix. They involve oxidation treatments and the use of coupling agents, wetting agents, and/or sizings (coatings). Carbon fibers need treatment for both thermosets and thermoplastics. As the processing temperature is usually higher for thermoplastics than thermosets, the treatment must be stable to a higher temperature (300–400°C) when a thermoplastic is used.

Oxidation treatments can be applied by gaseous, solution, electrochemical, and plasma methods. Oxidizing plasmas include those involving oxygen,[5-8] CO_2,[9] and air.[5,10] The resulting oxygen-containing functional groups (hydroxyl, ketone, and carboxyl groups) on the fiber surface cause improvement in the wettability of the fiber, and in fiber-matrix adhesion. The consequence is enhancement of the interlaminar shear strength (ILSS) and flexural strength. Other plasmas (not necessarily oxidizing) that are effective involve nitrogen,[5] acrylonitrile,[5] and trimethyl silane.[11]

TABLE 8.1
Properties of Thermoplastics for Carbon Fiber Polymer-Matrix Composites

	PES	PEEK	PEI	PPS	PI
T_g (°C)	230[a]	170[a]	225[a]	86[a]	256[b]
Decomposition temperature (°C)	550[a]	590[a]	555[a]	527[a]	550[b]
Processing temperature (°C)	350[a]	380[a]	350[a]	316[a]	304[b]
Tensile strength (MPa)	84[c]	70[c]	105[c]	66[c]	138[b]
Modulus of elasticity (GPa)	2.4[c]	3.8[c]	3.0[c]	3.3[c]	3.4[b]
Ductility (% elongation)	30–80[c]	50–150[c]	50–65[c]	2[c]	5[b]
Izod impact (ft lb/in.)	1.6[c]	1.6[c]	1[c]	<0.5[c]	1.5[c]
Density (g/cm³)	1.37[c]	1.31[c]	1.27[c]	1.3[c]	1.37[b]

[a] From Ref. 1
[b] From Ref. 2
[c] From Ref. 3

Akin to plasma treatment is ion beam treatment, which involves oxygen or nitrogen ions.[12] Plasma treatments are useful for epoxy as well as thermoplastic matrices.[6,13-16] Oxidation by gaseous methods includes the use of oxygen gas containing ozone.[17] Oxidation by solution methods involves wet oxidation,[18,19] such as acid treatments.[6,20,21] Oxidation by electrochemical methods includes the use of ammonium sulfate solutions,[22] a diammonium hydrogen phosphate solution containing ammonium rhodanide,[23] ammonium bicarbonate solutions,[24,25] a phosphoric acid solution,[26] and other aqueous electrolytes.[27-30] In general, the various treatments provide chemical modification of the fiber surface in addition to removal of a loosely adherent surface layer.[31-36] More severe oxidation treatments serve to roughen the fiber surface, enhancing the mechanical interlocking between the fibers and the matrix.[37,38]

Table 8.2[39] shows the effect of gaseous and solution oxidation treatments on the mechanical properties of high-modulus carbon fibers and their epoxy-matrix composites. The treatments degrade the fiber properties but improve the composite properties. The most effective treatment in Table 8.2 is refluxing in a 10% $NaClO_3/25\%$ H_2SO_4 mixture for 15 min, as this treatment results in a fiber weight loss of 0.2%, a fiber tensile strength loss of 2%, a composite flexural strength gain of 5%, and a composite interlaminar shear strength (ILSS) gain of 91%. Epoxy-embedded single fiber tensile testing showed that anodic oxidation of pitch-based carbon fibers in ammonium sulfate solutions increased the interfacial shear strength by 300%.[40] As the modulus of the fiber increases, progressively longer treatment times are required to attain the same improvement in ILSS. Although the treatment increases ILSS, it decreases the impact strength, so treatment time must be carefully controlled in order to achieve a balance in properties. The choice of treatment time also depends on the particular fiber-resin combination used. For a particular treatment, as the modulus of the fiber increases, the treatment's positive effect on the ILSS and its negative effect on the impact strength become more severe.[39]

TABLE 8.2

Effects of Various Surface Treatments on Properties of High-Modulus Carbon Fibers and Their Epoxy-Matrix Composites (From Ref. 39)

| | Fiber Properties | | Composite Properties | |
Fiber Treatment	Wt. Loss (%)	Tensile Strength Loss (%)	Flexural Strength Loss (%)	ILSS Gain (%)
400°C in air (30 min)	0	0	0	18
500°C in air (30 min)	0.4	6	12	50
600°C in air (30 min)	4.5	50	Too weak to test	—
60% HNO_3 (15 min)	0.2	0	8	11
5.25% NaOCl (30 min)	0.4	1.5	5	30
10–15% NaOCl (15 min)	0.2	0	8	6
15% $HClO_4$ (15 min)	0.2	0	12	0
5% $KMnO_4$/10% NaOH (15 min)	0.4	0	15	19
5% $KMnO_4$/10% H_2SO_4 (15 min)	6.0(+)	17	13	95
10% H_2O_2/20% H_2SO_4 (15 min)	0.1	5	14	0
42% HNO_3/30% H_2SO_4 (15 min)	0.1	0	4(+)	0
10% $NaClO_3$/15% NaOH (15 min)	0.2	0	12	12
10% $NaClO_3$/25% H_2SO_4 (15 min)	0.2	2	5(+)	91
15% $NaClO_3$/40% H_2SO_4 (15 min)	0.7	4	15	108
10% $Na_2Cr_2O_7$/25% H_2SO_4 (15 min)	0.3	8	15(+)	18
15% $Na_2Cr_2O_7$/40% H_2SO_4 (15 min)	1.7	27	31	18

Note: All liquid treatments at reflux temperature.

Less common methods of modification of the carbon fiber surface involve using gamma-ray radiation for surface oxidation,[41] and applying coatings by electrochemical polymerization (polyphenylene oxide, PPO, as a coating),[42-45] plasma polymerization,[43,44,46-49] vapor deposition polymerization,[50] and other polymerization techniques.[51-54] A relatively simple coating technique is solution dipping,[43] as in the case of coating with polyurethane.[55] Tough, compliant thermoplastic coatings,[56] such as polyurethane,[55] silicone,[57,58] and polyvinyl alcohol[59] are attractive for enhancing the toughness of composites. Other thermoplastic coatings, such as polyetherimide (PEI), improve epoxy matrix properties through diffusion into the epoxy matrix.[60] Still other coatings serve as coupling agents between the fiber and the matrix because of the presence of reactive functional groups, such as epoxy groups[61] and amine groups.[62] Metal (e.g., nickel and copper) coatings are used not only for enhancing the electrical properties,[63,64] but also for improving the adhesion between fiber and the epoxy matrix.[65]

Commercial carbon fibers are surface treated to enhance the bonding with epoxy, though the surface treatment is proprietary. Table 8.3[66] shows the effect of a surface treatment on the interfacial shear strength for PAN-based carbon fibers manufactured by Hercules. Fibers designated AS-1 and AS-4 are typical Type II intermediate strain

TABLE 8.3
Interfacial Shear Strength of
Carbon Fibers in an Epoxy Matrix
(From Ref 66)

Fiber	Interfacial Shear Strength (MPa)
AU-1	48
AS-1	74
AU-4	37
AS-4	61

TABLE 8.4
Atomic Surface Concentrations
of the Carbon Fibers of Table 8.3
(From Ref. 66)

Fiber	Atomic Percent				
	C_{1s}	O_{1s}	N_{1s}	S_{2p}	Na_{KLL}
AS-1	84	11	4.3	0.2	1.0
AS-4	83	12	4.0	0.2	0.7

fibers, whereas AU-1 and AU-4 are the untreated analogs of AS-1 and AS-4, respectively. The interfacial shear strength was determined from the critical shear transfer length, i.e., the length of the fiber fragments after fracture of a single fiber pulled in tension while being encapsulated in epoxy. Surface treatment increased the interfacial shear strength by 54 and 65% for AS-1 and AS-4, respectively.[66]

Table 8.4[66] shows the atomic surface concentrations of AS-1 and AS-4 fibers, as determined by x-ray photoelectron spectroscopy. The atomic surface concentrations are similar for AS-1 and AS-4, indicating that the superior interfacial shear strength of AS-1 compared to AS-4 is not due to a difference in surface composition. On the other hand, scanning electron microscopy shows that the surface of AS-1 is corrugated, whereas that of AS-4 is smooth. The superior interfacial shear strength of AS-1 is attributed to its surface morphology, which increases its surface area and enhances the mechanical interlocking between the fiber and the matrix.[66]

Although surface treatments of carbon fibers result in some degree of oxidation, which places oxygen on the surface in an acidic form, the treatments themselves produce little acidity. Surface acidification is not desirable because it is accompanied by surface degradation.[67]

The oxidation treatments approximately double the surface concentration of oxygen. Functional groups on the fiber surface and at 500 Å below the surface are listed below in relative order of abundance.

At the surface

$$-\overset{|}{\underset{|}{C}}-O- \quad > \quad -\overset{|}{\underset{|}{C}}-OH \quad = \quad -\overset{|}{\underset{|}{C}}-S- \quad > \quad -\overset{O}{\overset{\|}{C}}-OH \quad = \quad -\overset{O}{\overset{\|}{C}}- \quad > \quad -\overset{|}{\underset{|}{C}}-Cl$$

At 500 Å depth

$$-\overset{|}{\underset{|}{C}}-O- \quad > \quad -\overset{|}{\underset{|}{C}}-S- \quad = \quad -\overset{O}{\overset{\|}{C}}- \quad > \quad -O-\overset{O}{\overset{\|}{C}}-$$

The main functional groups produced are carbonyl, carboxyl, and hydroxyl.[39]

The oxygen concentration does not simply determine the increase in ILSS or transverse tensile strength of the composites.[14,23,24,26,27,31,37,39,44] Indeed, the addition of the surface chemical oxygen groups is believed to be responsible for only 10% of the increase in adhesion resulting from the treatment; only about 4% of the surface sites of the carbon fibers are involved in chemical bonding with the epoxy and amine groups of the polymer. Although the magnitude of the bond strength for chemical bonds is very high, the quantity of bonds is low.[68] On the other hand, elimination of the functional groups on the treated fibers by diazomethane causes the ILSS to decrease toward the level of the untreated fibers,[69] so the contribution of the functional groups to fiber-matrix adhesion cannot be neglected.

In addition to oxidation treatments, carbon fibers require the use of coupling agents, wetting agents, and/or sizings in order to improve the wetting of the fibers by the polymer, the adhesion between the fibers and the matrix, and the handleability of the fibers. As one agent often serves more than one function, the distinction among coupling agents, wetting agents, and sizings is often vague.

Coupling agents[39,61,62,70] are mostly short-chain hydrocarbon molecules. One end of a molecule is compatible with or interacts with the polymer, while the other end interacts with the fiber. A coupling agent molecule has the form X–R, where X interacts with the fiber and R is compatible with or interacts with the polymer.

Organosilanes are of the form R–Si–$(OX)_3$, where X is methyl, ethyl, methoxyethyl, etc., and R is a suitable hydrocarbon chain. They are widely used as coupling agents between glass fibers and thermosets, as the –OX groups react with the –OH groups on the glass surface.

$$\begin{array}{ccc} | & & | \\ O & & O \\ | & & | \\ Si-OH + XO-R & \rightarrow & Si-O-R + HOX \\ | & & | \\ O & & O \\ | & & | \end{array}$$

However, organosilanes do not function for carbon, organic, or metallic fibers.

Organotitanates and organozirconates function as coupling agents for both sili-
ceous and nonsiliceous fillers. The general formula of the organotitanates is:

$$
X\text{-}O\text{-}Ti \left[\begin{array}{c} \\ O\text{-}P\text{-}O\text{-}P\text{-}(OR)_2 \\ \end{array} \right]_3
$$

where R is usually a short-chain hydrocarbon such as C_8H_{17} and X is a group capable
of interacting with the fiber. Organotitanates and organozirconates in amounts of
0.1–0.5 wt.% of formulation solids provide improved bonding between carbon fibers
and thermosets (epoxy, polyurethane, polyester, and vinyl ester resins).[71]

Wetting agents are polar molecules with one end attracted to the fiber and the
other to the polymer. The agent forms a protective layer around the fiber, thereby
improving dispersion. It also promotes adhesion by allowing more efficient wetting
of the polymer on the fiber. The main difference between a wetting agent and a
coupling agent is that a coupling agent forms a chemical bond with the fiber but a
wetting agent does not.[39]

Sizings are applied to carbon fibers to improve fiber-polymer adhesion and fiber
handleability. Handleability is particularly important if the fibers are to be woven.
The choice of sizing material depends on the polymer matrix. In particular, thermoplastic-matrix composites require sizings that can withstand higher temperatures
than thermoset-matrix composites because of the higher processing temperature of
the fiber. Sizing materials include prepolymers/polymers, carbon, SiC, and metals.
Because of the relative ease of application, polymers are the most common sizing
materials. Sizing thicknesses typically range from 0.1 to 1 µm.

Commercial carbon fibers are usually coated with a proprietary epoxy-compat-
ible finish. Nevertheless, epoxy is the main sizing material for fibers used for epoxy-
matrix composites. As the epoxy sizing decomposes at about 250°C, it is not very
suitable for thermoplastic-matrix composites, though it is still useful.[72] Instead,
polimides and polyimide-PES blends are used as sizings for carbon fibers in thermoplastic-matrix composites. Polyimide-coated carbon fibers can withstand temperatures up to 450°C.[1]

Other than epoxy, a number of polymers have been used as sizings for carbon
fibers in epoxy-matrix composites. They include polyhydroxyether, polyphenyle-
neoxide, copolymers of styrene and maleic anhydride (SMA), a block copolymer
of SMA with isoprene, polysulfone, polybutadiene, silicone, a carboxy-terminated
polybutadiene/acrylonitrile copolymer, a copolymer of maleic anhydride and buta-
diene, and a copolymer of ethylene and acrylic acid.[73] In particular, an SMA coating
results in a 50% increase in the interfacial shear strength compared with commercially treated fibers, while causing no degradation in impact strength.[74] In contrast,
elastomer coatings result in improved crack resistance and impact strength.[57,58,75]

Different methods are used to coat carbon fibers with polymers, namely deposition from solution, vapor, electrodeposition, and electropolymerization. Polymer
deposition from solution has been the most common, though recent work employs

mostly electrodeposition or electropolymerization. An example of deposition from solution is the deposition of polyhydroxyether from a solution containing 0.91 wt.% of polyhydroxyether in Cellosolve®. Polyhydroxyether is:

$$\left[O - Be - \underset{\underset{CH_3}{|}}{\overset{\overset{CH_3}{|}}{C}} - Be - O - CH_2 - \underset{\underset{OH}{|}}{\overset{\overset{H}{|}}{C}} - CH_2 \right]_n$$

The resulting polyhydroxyether coating increased the ILSS and flexural strength of the carbon fiber epoxy-matrix composite by 81% and 14%, respectively.[72] In electrodeposition, preformed polymers carrying ionized groups migrate to the oppositely charged electrode under an applied voltage. Electropolymerization involves the polymerization of monomers in an electolytic cell. Solvents such as dimethyl formamide and dimethyl sulfoxide have proved suitable. Since carbon fibers are electrically conducting, they serve as a good substrate for these electrical coating methods, which have the advantage of yielding uniform layers of readily controlled thickness in a short time.[76] Sometimes grafting of the polymer to the fiber surface takes place, as in the case of the electrochemical oxidation of ω-diamines on carbon fibers, where grafting provides a continuous succession of covalent bonds from the carbon fiber surface to the epoxy resin.[77] Deposition techniques have provided the greatest improvement in composite properties — up to 60% in impact strength, 84% in work to fracture, and 90% in ILSS, but not simultaneously.[73] Table 8.5[69] shows the effect of oxidation treatment and polymer coating on ILSS. The use of both oxidation treatment and polymer coating yields the highest ILSS.

Sizing increases the ILSS; this changes the mode of composite fracture from growth of an interfacial crack to growth of a crack perpendicular to the fiber axis.[39] In some cases, the epoxy matrix penetrates the polymer coating.[73] An interphase between the fiber and the epoxy matrix is believed to exist. It is a three-dimensional region including not only the two-dimensional fiber-matrix interface, but also regions on both sides of the interface.[68,78-80]

Polymer sizings are chemically bonded to carbon fibers via their functional groups. In the case of an epoxy sizing, the epoxy group and amine group can react with the functional groups on the fiber surface.[68] Furthermore, the functional groups act as catalysts for cross-linking if their concentrations are not too high.[76] In the case of a polyimide sizing, the carboxylic acid groups in polyimide precursors react with the functional groups on the fiber surface.[1]

Far less common than polymer coatings are carbon, SiC, and metal coatings on carbon fibers. Carbon coatings deposited by using acetylene vapor increase the ILSS from 34 MPa at 0 wt.% C coating to 56 MPa at 22 wt.% C coating.[69] SiC coatings, which can be in the form of β-SiC single-crystal whiskers grown on the carbon fiber surface perpendicular to the fiber axis, significantly increase the ILSS.[69,81] Metal (Ni, Cu) coatings deposited by electroless plating or electroplating on carbon fibers provide polar surfaces due to the presence of oxides and hydration of the surface.[65,82]

TABLE 8.5
Effect of Oxidation Treatment and Polymer Coating on
ILSS of Composite (From Ref. 69)

Oxidation	Polymer Coating	Polymer (%)	Density (g/cm^3)	ILSS (MPa)
None	None	—	1.28	16.2
60% HNO$_3$, 24 h.	None	—	1.29	24.3
	PVA	7	1.31	42.8
	PVC	7	1.31	42.1
	Rigid polyurethane	3	1.27	40.7
	PAN	7	1.27	16.6

8.4 SURFACE MODIFICATION OF CARBON FIBERS FOR METAL-MATRIX COMPOSITES

The difficulty molten metals have in wetting the surface of carbon fibers complicates the fabrication of metal-matrix composites. This difficulty is particularly severe for high-modulus carbon fibers (e.g., BP Amoco's Thornel® P-100) that have graphite planes mostly aligned parallel to the fiber surface. The edges of the graphite planes are more reactive with the molten metals than the graphite planes themselves. As a result, low-modulus carbon fibers are more reactive and are wetted more easily by the molten metals. Although this reaction between the fibers and the metal helps the wetting, it produces a brittle carbide and degrades the strength of the fibers.

To enhance wetting or other characteristics, carbon fibers are coated by a metal, a ceramic, or carbon. The metals include nickel[83,84] and silver.[85,86] Nickel coating can be attained by either electroplating or carbonyl decomposition. The ceramics include TiB$_2$,[87,88] SiC,[89-94] Al$_2$O$_3$,[95,96] TiC,[94] B$_4$C$_3$,[94] and ZrO$_2$.[97] The SiC coating can be produced by using a polycarbosilane solution,[90,91,93] which is subsequently pyrolyzed. However, it can also be produced by reactive chemical vapor deposition (RCVD) using SiCl$_4$ gas.[94] Zirconia (ZrO$_2$) coating can be produced from a zirconium oxychloride solution by dip-coating and heating.[97] The carbon coating is pyrolytic carbon[84,87-89] obtained by chemical vapor deposition.

A double layer coating comprising a carbon inner layer and an outer layer which is either TiB$_2$[87,88] or SiC[89] is particularly attractive because the carbon layer deviates cracks, while the ceramic layer enhances wettability with liquid metals. Moreover, TiB$_2$ protects the fiber from the reaction with liquid aluminum that forms brittle Al$_4$C$_3$. A three-layer coating[98] comprising an inner carbon layer, an outer silicon layer, and an intermediate gradient layer (gradient in Si$_x$C$_y$ composition ranging from pure C at the interface with the carbon inner layer to pure Si at the interface with the silicon outer layer) is also attractive. The gradient layer serves as a wetting agent, a diffusion/reaction barrier, and a releaser of residual thermal stress, and allows tailoring of the interfacial shear strength.

Another method involves treating the carbon fiber with a K_2ZrF_6 solution, which enhances the wettability with liquid aluminum because of the reaction between K_2ZrF_6 and Al.[99,100]

In the case of Mg as the metal matrix, SiO_2 is the most common coating, as it is air-stable and improves the wetting and bonding between fiber and liquid magnesium.[101] A three-layer coating involving carbon as the inner layer, SiO_2 as the outer layer, and Si_xC_y as the gradient intermediate layer is particularly attractive.[102,103] The SiO_2 coating can be obtained by passing the fibers through a toluene solution containing a silicon-based organometallic compound and chloride, followed by hydrolysis and pyrolysis of the organometallic compound.[101] Another ceramic coating used for Mg-matrix composites is TiN, formed by chemical vapor deposition from $TiCl_4$ and N_2 gases.[104]

Metals used as coatings include Ni, Cu, and Ag; they generally result in composites of strengths much lower than those predicted by the rule of mixtures (ROM). In the case of nickel-coated and copper-coated carbon fibers in an aluminum matrix, metal aluminides (Al_3M) form and embrittle the composites. In the case of nickel-coated carbon fibers in a magnesium matrix, nickel reacts with magnesium to form Ni/Mg compounds and a low-melting (508°C) eutectic.[105] On the other hand, copper-coated fibers are suitable with copper,[106-110] tin, or other metals as the matrix. A metal coating that is particularly successful involves sodium, which wets carbon fibers and coats them with a protective intermetallic compound by reaction with one or more other molten metals (e.g., tin). This is called the sodium process.[111,112] A related process immerses the fibers in liquid NaK.[113] However, these processes involving sodium suffer from sodium contamination of the fibers, probably due to the intercalation of sodium into graphite.[114] Nevertheless, aluminum-matrix composites containing unidirectional carbon fibers treated by the sodium process exhibit tensile strengths close to those calculated by using the rule of mixtures, indicating that the fibers are not degraded by the sodium process.[114]

Ceramics used as coatings on carbon fibers include TiC, SiC, B_4C, TiB_2, TiN, K_2ZrF_6, and ZrO_2. Methods used to deposit the ceramics include (1) reaction of the carbon fibers with a molten metal alloy, called the liquid metal transfer agent (LMTA) technique; (2) chemical vapor deposition (CVD); and (3) solution coating.

The LMTA technique involves immersing the fibers in a melt of copper or tin (called a liquid metal transfer agent, which must not react with carbon) in which a refractory element (e.g., W, Cr, Ti) is dissolved, and subsequent removal of the transfer agent from the fiber surface by immersion in liquid aluminum. For example, to form a TiC coating, the alloy can be Cu–10% Ti at 1050°C or Sn–1% Ti at 900–1055°C. In particular, by immersing the fibers in Sn–1% Ti at 900–1055°C for 0.25–10 min, a 0.1 μm layer of TiC is formed on the fibers, although they are also surrounded by the tin alloy. Subsequent immersion for 1 min in liquid aluminum causes the tin alloy to dissolve in the liquid aluminum.[115] The consequence is a wire preform suitable for fabricating aluminum-matrix composites. Other than titanium carbide, tungsten carbide and chromium carbide have been formed on carbon fibers by the LMTA technique.

The CVD technique has been used for forming coatings of TiB_2, TiC, SiC, B_4C, and TiN. The B_4C coating is formed by reactive CVD on carbon fibers, using a

BCl_3/H_2 mixture as the reactant.[116] The TiB_2 deposition uses $TiCl_4$ and BCl_3 gases, which are reduced by Zn vapor. The TiB_2 coating is particularly attractive because of exceptionally good wetting between TiB_2 and molten aluminum. During composite fabrication, the TiB_2 coating is displaced and dissolved in the matrix, while an oxide (γ-Al_2O_3 for a pure aluminum matrix, $MgAl_2O_4$ spinel for a 6061 aluminum matrix) is formed between the fiber and the matrix. The oxygen for the oxide formation comes from the sizing on the fibers; the sizing is not completely removed before processing.[117] The oxide layer serves as a diffusion barrier to aluminum, but allows diffusion of carbon, thereby limiting Al_4C_3 growth to the oxide-matrix interface.[118] Moreover, the oxide provides bonding between the fiber and the matrix. Because of the reaction at the interface between the coating and the fiber, the fiber strength is degraded after coating. To alleviate this problem, a layer of pyrolytic carbon is deposited between the fiber and the ceramic layer.[119] The CVD process involves high temperatures, e.g., 1200°C for SiC deposition using CH_3SiCl_4;[120] this high temperature degrades the carbon fibers. Another problem of the CVD process is the difficulty of obtaining a uniform coating around the circumference of each fiber. Moreover, it is expensive and causes the need to scrub and dispose most of the corrosive starting material, as most of the starting material does not react at all. The most serious problem with the TiB_2 coating is that it is not air stable. It cannot be exposed to air before immersion in the molten metal or wetting will not take place. This limits the shape of materials that can be fabricated, especially since the wire preforms are not very flexible.[117]

A high compliance (or a low modulus) is preferred for the coating in order to increase the interface strength. An increase in the interface (or interphase) strength results in an increase in the transverse strength. The modulus of SiC coatings can be varied by controlling the plasma voltage in plasma-assisted chemical vapor deposition (PACVD). Modulus values in a range from 19 to 285 GPa have been obtained in PACVD SiC, compared to a value of 448 GPa for CVD SiC. Unidirectional carbon fiber (Thornel® P-55) aluminum-matrix composites in which the fibers are coated with SiC exhibit an interfacial strength and a transverse strength that increase with decreasing modulus of the SiC coating.[121-123]

The most attractive coating technique developed to date is the solution coating method. In the case of using an organometallic solution, fibers are passed through a toluene solution containing an organometallic compound, followed by hydrolysis or pyrolysis of the organometallic compounds to form the coating. Thus, the fibers are passed sequentially through a furnace in which the sizing on the fibers is vaporized, followed by an ultrasonic bath containing an organometallic solution. The coated fibers are then passed through a chamber containing flowing steam in which the organometallic compound on the fiber surface is hydrolyzed to oxide, and finally through an argon atmosphere drying furnace in which any excess solvent or water is vaporized and any unhydrolyzed organometallic is pyrolyzed.[117] In contrast to the TiB_2 coatings, the SiO_2 coatings formed by organometallic solution coating are air stable.

The organometallic compounds used are alkoxides, in which metal atoms are bound to hydrocarbon groups by oxygen atoms. The general formula is $M(OR)_x$, where R is any hydrocarbon group and x is the oxidation state of the metal atom M. When exposed to water vapor, these alkoxides hydrolyze.[117]

$$M(OR)_x + \frac{x}{2}H_2O \rightarrow MO_{x/2} + xROH$$

For example, the alkoxide tetraethoxysilane (also called tetraethylorthosilicate) is hydrolyzed by water as follows:[117]

$$Si(OC_2H_5)_4 + 2H_2O \rightarrow SiO_2 + 4C_2H_5OH$$

Alkoxides can also be pyrolyzed to yield oxides, e.g., Reference 117:

$$Si(OC_2H_5)_4 \rightarrow SiO_2 + 2C_2H_5OH + 2C_2H_4$$

Most alkoxides can be dissolved in toluene. By controlling the solution concentration and the time and temperature of immersion, it is possible to control the uniformity and thickness of the resulting oxide coating. The thickness of the oxide coatings on the fibers varies from 700 to 1500 Å. The oxide is amorphous and contains carbon, which originates in the carbon fiber.

Liquid magnesium wets SiO_2-coated low-modulus carbon fibers (e.g., T-300) and infiltrates the fiber bundles due to reactions between the molten magnesium and the SiO_2 coating. The reactions include the following.[117]

$$2Mg + SiO_2 \rightarrow 2MgO + Si \qquad \Delta G^\circ_{670°C} = -76 \text{ kcal}$$

$$MgO + SiO_2 \rightarrow MgSiO_3 \qquad \Delta G^\circ_{670°C} = -23 \text{ kcal}$$

$$2Mg + 3SiO_2 \rightarrow 2MgSiO_3 + Si \qquad \Delta G^\circ_{670°C} = -122 \text{ kcal}$$

$$2MgO + SiO_2 \rightarrow Mg_2SiO_4 \qquad \Delta G^\circ_{670°C} = -28 \text{ kcal}$$

$$2Mg + 2SiO_2 \rightarrow Mg_2SiO_4 + Si \qquad \Delta G^\circ_{670°C} = -104 \text{ kcal}$$

The interfacial layer between the fiber and the Mg matrix contains MgO and magnesium silicates. However, immersion of SiO_2-coated high-modulus fibers (e.g., P-100) in liquid magnesium causes the oxide coating to separate from the fibers because of the poor adherence of the oxide coating to the high-modulus fibers. This problem can be solved by first depositing a thin amorphous carbon coating on the fibers by passing the fiber bundles through a toluene solution of petroleum pitch, followed by evaporation of the solvent and pyrolysis of the pitch.

The most effective air-stable coating for carbon fibers used in aluminum-matrix composites is a mixed boron-silicon oxide applied from organometallic solutions.[117]

Instead of SiO_2, TiO_2 can be deposited on carbon fibers by the organometallic solution method. For TiO_2, the alkoxide can be titanium isopropoxide.[124]

SiC coatings can be formed by using polycarbosilane (dissolved in toluene) as the precursor, which is pyrolyzed to SiC. These coatings are wet by molten copper containing a small amount of titanium because of a reaction between SiC and Ti to form TiC.[117]

Instead of using an organometallic solution, another method uses an aqueous solution of a salt. For example, the salt, potassium zirconium hexafluoride (K_2ZrF_6) or potassium titanium hexafluoride (K_2TiF_6), is used to deposit microcrystals of K_2ZrF_6 or K_2TiF_6 on the fiber surface.[117,125,126] These fluoride coatings are stable in air. The following reactions supposedly take place[125] between K_2ZrF_6 and the aluminum matrix:

$$3K_2ZrF_6 + 4Al \rightarrow 6KF + 4AlF_3 + 3Zr \tag{1}$$

$$3Zr + 9Al \rightarrow 3Al_3Zr \tag{2}$$

$$Zr + O_2 \rightarrow ZrO_2 \tag{3}$$

In the case of an Al–12 wt.% Si alloy (rather than pure Al) as the matrix, the following reaction may also occur:[126]

$$Zr + 2Si \rightarrow ZrSi_2 \tag{4}$$

The fluorides KF and AlF_3 are thought to dissolve the thin layer of Al_2O_3 on the liquid aluminum surface, thus helping the liquid aluminum to wet the carbon fibers. Furthermore, reactions (1) and (2) are strongly exothermic and may cause a local temperature increase near the fiber-matrix interface. The increased temperature probably gives rise to a liquid phase at the fiber-matrix interface.[125]

Although the K_2ZrF_6 treatment causes the contact angle between carbon and liquid aluminum at 700–800°C to decrease from 160° to 60–75°C,[78] it causes degradation of the fiber tensile strength during aluminum infiltration.[127]

Another example of a salt solution coating method involves the use of zirconium oxychloride ($ZrOCl_2$).[128] Dip-coating the carbon fibers in the salt solution and subsequently heating at 330°C cause the formation of a ZrO_2 coating of less than 1 μm in thickness. The ZrO_2 coating improves fiber-matrix wetting and reduces the fiber-matrix reaction in aluminum-matrix composites.

Instead of treating the carbon fibers, the wetting of the carbon fibers by molten metals can be improved by the addition of alloying elements into the molten metals. For aluminum as the matrix, effective alloying elements include Mg, Cu, and Fe.[114]

REFERENCES

1. W.-T. Whang and W.-L. Liu, *SAMPE Q.*, 22(1), 3-9 (1990).
2. D.C. Sherman, C.-Y. Chen, and J.L. Cercena, *Proc. Int. SAMPE Symp. and Exhib.*, 33, G. Carrillo, E.D. Newell, W.D. Brown, and P. Phelan, Eds., pp. 538-539 (1988).
3. D.R. Askeland, *The Science and Engineering of Materials*, 2nd ed., PWS-Kent, Boston, pp. 538-539 (1989).
4. S.D. Mills, D.M. Lee, A.Y. Lou, D.F. Register, and M.L. Stone, Proc. 20th Int. SAMPE Tech. Conf., pp. 263-270 (1988).
5. G. Bogoeva-Gaceva, E. Maeder, L. Hauessler and A. Dekanski, *Composites — Part A: Applied Science and Manufacturing*, 28(5), 445-452 (1997).

6. J. Jang and H. Yang, *J. Mater. Sci.*, 35(9), 2297-2303 (2000).
7. G. Wu, C.-H. Hung and J.-C. Lu, International SAMPE Symposium & Exhibition, 44(I), 1090-1097 (1999).
8. H. Zhuang and J.P. Wightman, *J. Adhesion*, 62(1-4), 213-245 (1997).
9. R.E. Allred and W.C. Schimpf, *J. Adhesion Sci. Tech.*, 8(4), 383-394 (1994).
10. G.J. Farrow, K.E. Atkinson, N. Fluck and C. Jones, *Surf. Interface Anal.*, 23(5), 313-318 (1995).
11. E.A. Friis, B. Kumar, F.W. Cooke and H.K. Yasuda, Proc. 1996 5th World Biomaterials Congress, Vol. 1, p. 913 (1996).
12. S.-S. Lin and P.W. Yip, Proc. of the 1993 Fall Meeting, Interface Control of Electrical, Chemical, and Mechanical Properties Materials Research Society Symposium Proceedings, Vol. 318, pp. 381-386 (1994).
13. N. Dilsiz, *J. Adhesion Sci. Tech.*, 14(7), 975-987 (2000).
14. M.C. Paiva, C.A. Bernardo and M. Nardin, *Carbon*, 38(9), 1323-1337 (2000).
15. G. Akovali and N. Dilsiz, *Polym. Eng. Sci.*, 36(8), 1081-1086 (1996).
16. N. Chand, E. Schulz and G. Hinrichsen, *J. Mater. Sci. Lett.*, 15(15), 1374-1375 (1996).
17. T. Yoshikawa and A. Kojima, Quarterly Report of RTRI (Railway Technical Research Institute) (Japan), 32(3), 190-199 (1991).
18. P.W.M. Peters and H. Albertsen, *J. Mater. Sci.*, 28(4), 1059-1066 (1993).
19. H. Albertsen, J. Ivens, P. Peters, M. Wevers and I. Verpoest, *Compos. Sci. Tech.*, 54(2), 133-145 (1995).
20. L. Ibarra, A. Macias and E. Palma, *J. Applied Polym. Sci.*, 61(13), 2447-2454 (1996).
21. D. Cho, *J. Mater. Sci. Lett.*, 15(20), 1786-1788 (1996).
22. T.R. King, D.F. Adams and D.A. Buttry, *Composites*, 22(5), 380-387 (1991).
23. E. Fitzer, N. Popovska and H.-P. Rensch, *J. Adhesion*, 36(2-3), 139-149 (1991).
24. M.R. Alexander and F.R. Jones, *Surf. Interface Anal.*, 22(1), 230-235 (1994).
25. Fukunaga and S. Ueda, *Compos. Sci. Tech.*, 60(2), 249-254 (2000).
26. S.-J. Park and M.-H. Kim, *J. Mater. Sci.*, 35(8), 1901-1905 (2000).
27. Jones, *Surf. Interface Anal.*, 20(5), 357-367 (1993).
28. C.A. Baillie and M.G. Bader, *J. Mater. Sci.*, 29(14), 3822-3836 (1994).
29. J.A. Hrivnak and R.L. McCullough, *J. Thermoplastic Compos. Mater.*, 9(4), 304-315 (1996).
30. I.A. Rashkovan and Y.G. Korabel'nikov, *Compos. Sci. Tech.*, 57(8), 1017-1022 (1997).
31. G. Bogoeva-Gaceva, D. Burevski, A. Dekanski and A. Janevski, *J. Mater. Sci.*, 30(13), 3543-3546 (1995).
32. D.W. Dwight, Proc. 53rd Annual Technical Conf. — ANTEC, Vol. 2, pp. 2744-2747 (1995).
33. N. Tsujioka, Z. Maekawa, H. Hamada and M. Hojo, *Zairyo/J. Soc. Mater. Sci.* (Japan), 46(2), 163-169 (1997).
34. L. Ibarra and D. Panos, *Polym. Int.*, 43(3), 251-259 (1997).
35. L. Ibarra and D. Panos, *J. Applied Polym. Sci.*, 67(10), 1819-1826 (1998).
36. J.M.M. de Kok and T. Peijs, *Composites — Part A: Applied Science and Manufacturing*, 30(7), 917-932 (1999).
37. P.W. Yip and S.S. Lin, *Mater. Res. Soc. Symp. Proc.*, Vol. 170, C.G. Pantano and E.J.H. Chen, Eds., pp. 339-344 (1990).
38. T.C. Chang and B.Z. Jang, *Mater. Res. Soc. Symp. Proc.*, Vol. 170, C.G. Pantano and E.J.H. Chen, Eds., pp. 321-326 (1990).
39. W.W. Wright, *Compos. Polym.*, 3(4), 231-257 (1990).
40. T.R. King, D.F. Adams, and D.A. Buttry, *Composites*, 22(5), 380-387 (1991).

41. J. Tian, Q. Wang, S. Yang and Q. Xue, *Appl. Math. Mech.* (English Edition), 19(10), 1-5 (1998).
42. J.A. King, D.A. Buttry and D.F. Adams, *Polym. Compos.*, 14(4), 301-307 (1993).
43. B. Harris, O.G. Braddell, D.P. Almond, C. Lefebvre and J. Verbist, *J. Mater. Sci.*, 28(12), 3353-3366 (1993).
44. B. Harris, O.G. Braddell, C. Lefebvre and J. Verbist, *Plast. Rubber Compos.*, 18(4), 221-240 (1992).
45. F. Zhang and L. Hu, *Fuhe Cailiao Xuebao/Acta Materiae Compositae Sinica*, 14(2), 12-16 (1997).
46. A.P. Kettle, F.R. Jones, M.R. Alexander, R.D. Short, M. Stollenwerk, J. Zabold, W. Michaeli, W. Wu, E. Jacobs and I. Verpoest, *Composites — Part A: Applied Science and Manufacturing*, 29(3), 241-250 (1998).
47. N. Dilsiz and G. Akovali, *Composite Interfaces*, 3(5-6), 401-410 (1996).
48. N. Dilsiz, E. Ebert, W. Weisweiler and G. Akovali, *J. Colloid Interface Sci.*, 170(1), 241-248 (1995).
49. R. Li, L. Ye and Y.-W. Mai, *Composites — Part A: Applied Science and Manufacturing*, 28(1), 73-86 (1997).
50. J.P. Armistead and A.W. Snow, *ASTM Special Technical Publication*, (1290), 168-181 (1996).
51. T. Duvis, C.D. Papaspyrides and T. Skourlis, *Compos. Sci. Tech.*, 48(1-4), 127-133 (1993).
52. T. Skourlis, T. Duvis and C.D. Papaspyrides, *Compos. Sci. Tech.*, 48(1-4), 119-125 (1993).
53. Zheng, X. Wu and S. Li, *Huadong Huagong Xueyuan Xuebao/J. East China Inst. of Chem. Tech.*, 20(4), 485-491 (1994).
54. P.C. Varelidis, R.L. McCullough and C.D. Papaspyrides, *Compos. Sci. Tech.*, 59(12), 1813-1823 (1999).
55. M. Tanoglu, G.R. Palmese, S.H. McKnight and J.W. Gillespie, Jr., Proc. 1998 56th Annual Tech. Conf. — ANTEC, Vol. 2, pp. 2346-2350 (1998).
56. V. Giurgiutiu, K.L. Reifsnider, R.D. Kriz, B.K. Ahn J.J. Lesko, Proc. 36th AIAA/ASME/ASCE/AHS/ASC Structures, Structural Dynamics and Mater. Conf. and AIAA/ASME Adaptive Structures Forum, Vol. 1, pp. 453-469 (1995).
57. M. Labronici and H. Ishida, *Compos. Interfaces*, 5(3), 257-275 (1998).
58. M. Labronici and H. Ishida, *Compos. Interfaces*, 5(2), 87-116 (1998).
59. J.-K. Kim and Y.-W. Mai, Proc. Int. Conf. on Adv. Composite Mater., Minerals, Metals & Materials Soc. (TMS), pp. 69-77 (1993).
60. S. Shin and J. Jang, *J. Mater. Sci.*, 35(8), 2047-2054 (2000).
61. J. Gulyas, S. Rosenberger, E. Foldes and B. Pukanszky, *Polym. Compos.*, 21(3), 387-395 (2000).
62. J.A. King, D.A. Buttry and D.F. Adams, *Polym. Compos.*, 14(4), 292-300 (1993).
63. G. Lu, X. Li and H. Jiang, *Compos. Sci. Tech.*, 56(2), 193-2000 (1996).
64. F.A. Hussain and A.M. Zihlif, *J. Thermoplastic Compos. Mater.*, 6(2), 120-129 (1993).
65. Hage Jr., S.F. Costa and L.A. Pessan, *J. Adhesion Sci. Tech.*, 11(12), 1491-1499 (1997).
66. M.J. Rich and L.T. Drzal, *J. Reinf. Plast. Compos.*, 7(2), 145-154 (1988).
67. B.-W. Chun, C.R. Davis, Q. He and R.R. Gustafson, *Carbon*, 30(2), 177-187 (1992).
68. L.T. Drzal, *Vacuum*, 41(7-9), 1615-1618 (1990).
69. R. Yosomiya, K. Morimoto, A. Nakajima, Y. Ikada, and T. Suzuki, Eds., *Adhesion and Bonding in Composites*, Marcel Dekker, New York, pp. 257-281 (1990).

70. R. Yosomiya, K. Morimoto, A. Nakajima, Y. Ikada, and T. Suzuki, Eds., *Adhesion and Bonding in Composites,* Marcel Dekker, New York, pp. 109-154 (1990).

71. G. Sugerman, S.M. Gabayson, W.E. Chitwood, and S.J. Monte, *Proc. 3rd Dev. Sci. Technol. Compos. Mater., Eur. Conf. Compos. Mater.,* A.R. Bunsell, P. Lamicq, and A. Massiah, Eds., Elsevier, London, pp. 51-56 (1989).

72. K.T. Kern, E.R. Long, Jr., S.A.T. Long, and W.L. Harries, *Polym. Prepr.,* 31(1), 611-612 (1990).

73. W.W. Wright, *Compos. Polym.,* 3(5), 360-401 (1990).

74. R.V. Subramanian, A.R. Sanadi, and A. Crasto, *J. Adhes. Sci. Technol.,* 4(10), 329-346 (1990).

75. S.H. Jao and F.J. McGarry, *Proc. Int. SAMPE Tech. Conf. 22,* L.D. Michelove, R.P. Caruso, P. Adams, and W.H. Fossey, Jr., Eds., pp. 455-469 (1990).

76. Sellitti, J.L. Koenig, and H. Ishida, *Mater. Sci. Eng.,* A126, 235-244 (1990).

77. Barbier, J. Pinson, G. Desarmot, and M. Sanchez, *J. Electrochem. Soc.,* 137(6), 1757-1764 (1990).

78. L.T. Drzal, *Mater. Sci. Eng.,* A126, 289-293 (1990).

79. L.T. Drzal, *Mater. Res. Soc. Symp. Proc.,* Vol. 170, C.G. Pantano and E.J.H. Chen, Eds., pp. 275-283 (1990).

80. L.T. Drzal, *Treatise on Adhesion and Adhesives,* Vol. 6, R.L. Patrick, Ed., Marcel Dekker, New York, pp. 187-211 (1989).

81. M.G. Harwell, D.E. Hirt, D.D. Edie, N. Popovska and G. Emig, *Carbon,* 38(8), 1111-1121 (2000).

82. N.C.W. Judd, *Br. Polym. J.,* 9(4), 272-277 (1977).

83. T.F. Stephenson, J.A.E. Bell and J.R. Gordon, 3rd Int. SAMPE Metals and Metals Processing Conf. and 24th Int. SAMPE Tech. Conf., Vol. 3, pp. 560-568 (1992).

84. B. Wielage and A. Dorner, *Compos. Sci. Tech.,* 59(8), 1239-1245 (1999).

85. S.G. Warrier and R.Y. Lin, *J. Mater. Sci.,* 28(18), 4868-4877 (1993).

86. S.G. Warrier, C.A. Blue and R.Y. Lin, *J. Mater. Sci.,* 28(3), 760-768 (1993).

87. P. Bertrand, M.H. Vidal-Setif, R. Valle and R. Mevrel, *J. Mater. Sci.,* 33(20), 5029-5036 (1998).

88. P. Bertrand, M.H. Vidal-Setif and R. Mevrel, *Surf. Coat. Tech.,* 96(2-3), 283-292 (1997).

89. O. Perez, G. Patriarche, M. Lancin and M.H. Vidal-Setif, *J. de Physique I.,* 3(7), Pt. 3, 1693-1698 (1993).

90. Y.-Q. Wang and B.-L. Zhou, *Composites — Part A: Applied Science and Manufacturing,* 27(12), 1139-1145 (1996).

91. Y.-Q. Wang and B.-L. Zhou, *J. Mater. Process. Tech.,* 73(1-3), 78-81 (1998).

92. J. Wang, T. Hong, G. Li and P. Li, *Composites — Part A: Applied Science and Manufacturing,* 28(11), 943-948 (1997).

93. Y.-Q. Wang, Z.-M. Wang, J.-Y. Yang, F.-Q. Zhang, B.-L. Zhou, *Compos. Manufacturing,* 6(2), 103-106 (1995).

94. H. Vincent, C. Vincent, M.P. Berthet, H. Mourichoux and J. Bouix, *J. Less-Common Met.,* 175(1), 37-58 (1991).

95. Q. Zeng, *J. Applied Polym. Sci.,* 70(1), 177-183 (1998).

96. Y. Wang, J. Zheng, Z. Wang, B. Zhou and L. Zhou, *Fuhe Cailiao Xuebao/Acta Materiae Compositae Sinica,* 12(3), 31-35 (1995).

97. R.V. Subramanian and E.A. Nyberg, *J. Mater. Res.,* 7(3), 677-688 (1992).

98. J.K. Yu, H.L. Li and B.L. Shang, *J. Mater. Sci.,* 29(10), 2641-2647 (1994).

99. G. Zheng, X. Chen, Z. Shen and H. Du, *Fuhe Cailiao Xuebao/Acta Materiae Compositae Sinica,* 10(1), 57-63 (1993).

100. X. Chen, G. Zhen and Z. Shen, *J. Mater. Sci.*, 31(16), 4297-4302 (1996).
101. R. Chen and X. Li, *Compos. Sci. Tech.*, 49(4), 357-362 (1993).
102. K. Zhang, Y. Wang and B. Zhou, *Transactions of Nonferrous Metals Soc. of China*, 7(3), 86-89 (1997).
103. K. Zhang, Y.Q. Wang, B.L. Zhou, Y.H. Zhou and H.L. Li, *Acta Metallurgica Sinica* (English Letters), 10(5), 398-402 (1997).
104. N. Popovska, H. Gerhard, D. Wurm, S. Poscher, G. Emig and R.F. Singer, *Mater. Design*, 18(4-6), 239-242 (1997).
105. I.W. Hall, *Metallography*, 20(2), 237-246 (1987).
106. D.A. Foster, *Proc. Int. SAMPE Symp. and Exhib.*, 34, G.A. Zakrzewski, D. Mazenko, S.T. Peters, and C.D. Dean, Eds., pp. 1401-1410 (1989).
107. P. Sebo, *Key Eng. Mater.*, 108-110, 315-322 (1995).
108. Y.Z. Wan, Y.L. Wang, G.J. Li, H.L. Luo and G.X. Cheng, *J. Mater. Sci. Lett.*, 16(19), 1561-1563 (1997).
109. M. Jahazi and F. Jalilian, *Compos. Sci. Tech.*, 59(13), 1969-1975 (1999).
110. S. Sun and M. Zhang, *Acta Metallurgica Sinica*, 26(6), B433-B437 (1990).
111. M.F. Amateau, *J. Compos. Mater.*, 10, 279 (1976).
112. D.M. Goddard, *J. Mater. Sci.*, 13(9), 1841-1848 (1978).
113. A.P. Levitt and H.E. Band, U.S. Patent 4,157,409 (1979).
114. R. Yosomiya, K. Morimoto, A. Nakajima, Y. Ikada, and T. Suzuki, Eds., *Adhesion and Bonding in Composites*, Marcel Dekker, New York, pp. 235-256 (1990).
115. D.D. Himbeault, R.A. Varin, and K. Piekarski, *Proc. Int. Symp. Process. Ceram. Met. Matrix Compos.*, H. Monstaghaci, Ed., Pergamon, New York, pp. 312-323 (1989).
116. H. Vincent, C. Vincent, J.P. Scharff, H. Mourichoux, and J. Bouix, *Carbon*, 30(3), 495-505 (1992).
117. H. Katzman, *Proc. Metal and Ceramic Matrix Composite Processing Conf.*, Vol. I, U.S. Dept. of Defense Information Analysis Centers, pp. 115-140 (1984); *J. Mater. Sci.* 22, 144-148 (1987); *Mater. Manufact. Process.*, 5(1), 1-15 (1990).
118. L.D. Brown and H.L. Marcus, *Proc. Metal and Ceramic Matrix Composite Processing Conf.*, Vol. II, U.S. Dept. of Defense Information Analysis Centers, pp. 91-113 (1984).
119. G. Leonhardt, E. Kieselstein, H. Podlesak, E. Than, and A. Hofmann, *Mater. Sci. Eng.*, A135, 157-160 (1991).
120. K. Honjo and A. Shindo, *Proc. 1st Compos. Interfaces Int. Conf.*, H. Ishida and J.L. Koenig, Eds., pp. 101-107 (1986).
121. J.A. Cornie, A.S. Argon, and V. Gupta, *MRS Bull.*, 16(4), 32-38 (1991).
122. H. Landis, Ph.D. dissertation, MIT, 1988.
123. A.S. Argon, V. Gupta, K.S. Landis, and J.A. Cornie, *J. Mater. Sci.*, 24, 1207-1218 (1989).
124. J.P. Clement and H.J. Rack, *Proc. Am. Soc. Compos. Symp. High Temp. Compos.*, pp. 11-20 (1989).
125. S. Schamm, J.P. Rocher, and R. Naslain, *Proc. 3rd Eur. Conf. Compos. Mater., Dev. Sci. Technol. Compos. Mater.*, A.R. Bunsell, P. Lamicq, and A. Massiah, Eds., Elsevier, London, pp. 157-163 (1989).
126. S.N. Patankar, V. Gopinathan, and P. Ramakrishnan, *Scripta Metall.*, 24, 2197-2202 (1990).
127. S.N. Patankar, V. Gopinathan, and P. Ramakrishnan, *J. Mater. Sci. Lett.*, 9, 912-913 (1990).
128. R.V. Subramanian and E.A. Nyberg, *J. Mater. Res.*, 7(3), 677-688 (1992).

9 Corrosion Control of Steel-Reinforced Concrete

CONTENTS

SYNOPSIS   The methods and materials for corrosion control of steel-reinforced concrete are reviewed. The methods are steel surface treatment, the use of admixtures in concrete, surface coating on concrete, and cathodic protection.

RELEVANT APPENDIX: *H*

9.1 INTRODUCTION

Steel-reinforced concrete is widely used in construction. The corrosion of the steel reinforcing bars (rebars) in concrete limits the life of concrete structures. It is one of the main causes for the deterioration of the civil infrastructure. Corrosion occurs in the steel regardless of the inherent capacity of concrete to protect the steel from corrosion; accelerated corrosion results from the loss of alkalinity in the concrete or the penetration of aggressive ions (such as chloride ions).

Methods of corrosion control of steel-reinforced concrete include cathodic protection,[1-12] surface treatments of the rebars (epoxy coating,[13-44] galvanizing,[21,32,45-51] copper cladding,[52] protective rust growth,[53] surface oxidation,[54] and sandblasting[54]),

131

the use of admixtures (organic and inorganic corrosion inhibitors,[51,55-69] silica fume,[71-89] fly ash,[90-92] slag,[93] and latex[70,94-96]) and the use of surface coating.[97-99] This chapter is a review of the methods and materials for corrosion control of steel-reinforced concrete.

9.2 STEEL SURFACE TREATMENT

Steel rebars are made of mild steel because of low cost. (Stainless steel is excellent in corrosion resistance,[100] but its high cost makes it impractical for use in concrete.) The coating of a steel rebar with epoxy is commonly used to improve corrosion resistance, but it degrades the bond between rebar and concrete, and the tendency of the epoxy coating to debond is a problem.[13-44] Furthermore, the cut ends of the rebar and areas of the rebar where the epoxy coating is damaged are not protected from corrosion. On the other hand, galvanized steel attains corrosion protection by its zinc coating, which acts as a sacrificial anode. Galvanized steel bonds to concrete better than epoxy-coated steel,[48] and the tendency of the coating to debond is also less. Areas of the rebar where the zinc coating is damaged are still protected; the exposed areas, such as the cut ends, are protected provided they are less than 8 mm from the zinc coating.[49] Steel surface treatments that improve both corrosion resistance and bond strength are attractive. They include sandblasting and surface oxidation.[54]

Sandblasting involves discharging ceramic particles (typically alumina particles around 250 μm) under pressure (about 80 psi or 0.6 MPa). It results in roughening as well as cleaning the surface of the steel rebar. Cleaning relates to the removal of rust and other contaminants on the surface, as a steel rebar is usually covered by rust and other contaminants. Cleaning causes the surface of the rebar to be more uniform in composition, which improves corrosion resistance. Roughening enhances the mechanical interlocking between rebar and concrete, thus increasing bond strength.[54]

Water immersion means total immersion of the rebar in water at room temperature for two days. It causes the formation of a black oxide layer on the surface of the rebar, thus enhancing the composition uniformity of the surface and improving corrosion resistance. In addition, the oxide layer enhances the adhesion between rebar and concrete, thereby increasing bond strength. Water immersion times that are less than or greater than two days yield less desirable effects on both bond strength and corrosion resistance.[54]

Steel rebars can also be coated with a corrosion-inhibiting cement slurry[51,55,101] or a cement-polymer composite[101] for corrosion protection.

Of all the methods described for treating the surface of steel rebar, the most widely used are epoxy coating and galvanizing because of their long history of usage.

9.3 ADMIXTURES IN CONCRETE

Admixtures are solids or liquids that are added to a concrete mix to improve the properties of the resulting concrete. Admixtures that enhance the corrosion resistance of steel reinforced concrete include those that are primarily for corrosion inhibition

TABLE 9.1

Effect of Carbon Fibers (f), Methylcellulose (M), Silica Fume (SF), and Latex (L) on the Corrosion Resistance of Steel Rebar in Concrete

	In saturated Ca(OH)$_2$ solution		In 0.5 N NaCl solution	
	E_{corr}*(−mV, ±5)	I_{corr}*(μA/cm², ±0.03)	E_{corr}*(−mV, ±5)	I_{corr}*(μA/cm², ±0.03)
P	210	0.74	510	1.50
+ M	220	0.73	/	/
+ M + f	220	0.68	560	2.50
+ M + SF	137	0.17	/	/
+ M + f + SF	170	0.22	350	1.15
+ SF	140	0.19	270	0.88
+ L	180	0.36	360	1.05
+ L + f	190	0.44	405	1.28

Note: P = plain, M = methylcellulose, f = carbon fibers, SF = silica fume, L = latex

* Value at 25 weeks of corrosion testing

and those that are primarily for improving the structural properties. The latter are attractive because of multifunctionality. The former are mostly inorganic chemicals (such as calcium nitrite,[56,67-70] copper oxide,[59] zinc oxide,[59] sodium thiocyanate,[60,61] and alkaline earth silicate[64]) that increase the alkalinity of the concrete, although they can be organic chemicals such as banana juice.[62] Admixtures primarily for structural property improvement can be solid particles such as silica fume,[71-89] fly ash[90-92] and slag,[93] and solid particle dispersions such as latex.[70,94-96]

Silica fume as an admixture is particularly effective for improving the corrosion resistance of steel-reinforced concrete due to the decrease in the water absorptivity, and not so much because of the increase in electrical resistivity.[71-89] Latex improves corrosion resistance because it decreases water absorptivity and increases electrical resistivity.[70,94-96] Methylcellulose improves corrosion resistance only slightly.[72] Carbon fibers decrease corrosion resistance due to a decrease in electrical resistivity.[72] However, the negative effect of the carbon fibers can be compensated by adding either silica fume or latex, which reduce water absorptivity.[72] The corrosion resistance of carbon fiber-reinforced concrete, which typically contains silica fume for improving fiber dispersion, is superior to that of plain concrete.[72]

Table 9.1[72] shows the effects of silica fume, latex, methylcellulose, and short carbon fibers as admixtures on the corrosion potential (E_{corr}, measured according to ASTM C876 using a high-impedance voltmeter and a saturated calomel electrode placed on the concrete surface; E_{corr} that is more negative than −270 mV suggests 90% probability of active corrosion) and the corrosion current density (I_{corr}, determined by measuring the polarization resistance at a low scan rate of 0.167 mV/s) of steel-reinforced concrete in both saturated Ca(OH)$_2$ and 0.5 N NaCl solutions. The saturated Ca(OH)$_2$ solution simulates the ordinary concrete environment; the NaCl solution represents a high-chloride environment. Silica fume improves the corrosion resistance of rebars in concrete in both saturated $Ca(OH)_2$ and NaCl solutions more effectively than any of the other admixtures, although latex is effective. Methylcellulose slightly improves the corrosion resistance of rebar in concrete in $Ca(OH)_2$ solution. Carbon fibers decrease the corrosion resistance of rebars in concrete, mainly because they decrease the electrical resistivity of concrete. The negative effect of fibers can be compensated by either silica fume or latex.

Instead of using a corrosion-inhibiting admixture in the entire volume of concrete, one may use the admixture to modify the cement slurry that is used as a coating on the steel rebar.[51,55] Compared to the use of rebars that have been either epoxy coated or galvanized, this method suffers from its labor-intensive site-oriented process.[101] On the other hand, the use of a shop-coating based on a cement-polymer composite is an emerging alternative.[101]

Of all the admixtures described for improving the corrosion resistance of steel-reinforced concrete, the most widely used are calcium nitrite, silica fume, and latex.

9.4 SURFACE COATING ON CONCRETE

Coatings (such as acrylic rubber) can be applied to the concrete surface for the purpose of corrosion control through improving impermeability.[97-99] However, this method suffers from the poor durability of the coating, and the loss of corrosion protection in areas where the coating is damaged.

9.5 CATHODIC PROTECTION

Cathodic protection is an effective method for corrosion control of steel-reinforced concrete.[1-11] It involves the application of a voltage to force electrons to go to the steel rebar, thereby making the steel a cathode. As the voltage needs to be constantly applied, the electrical energy consumption is substantial. This can be alleviated by the use of carbon fiber-reinforced concrete.

As the steel rebar is embedded in concrete, the electrons need to go through the concrete in order to reach the rebar. However, concrete is not very conducting electrically. The use of carbon fiber-reinforced concrete for embedding the rebar facilitates cathodic protection, as the short carbon fibers enhance the conductivity of the concrete.[5]

For directing electrons to the steel-reinforced concrete, an electrical contact that is connected to the voltage supply is needed on the concrete. One choice of an electrical contact material is zinc, a coating deposited on the concrete by thermal spraying. It has a very low volume resistivity (thus requiring no metal mesh embedment), but it suffers from poor wear and corrosion resistance, the tendency to oxidize, high thermal expansion coefficient, and high material and processing costs. Another choice is a conductor-filled polymer,[12] that can be applied as a coating without heating, but it suffers from poor wear resistance, higher thermal expansion coefficient, and high material cost. Yet another choice is a metal (e.g., titanium) strip or wire embedded at one end in cement mortar that is in the form of a coating on the steel-reinforced concrete. The use of carbon fiber-reinforced mortar for this coating facilitates cathodic protection, as it is advantageous to enhance its conductivity.[5]

Due to the decrease in volume electrical resistivity associated with carbon fiber addition (0.35 vol. %) to concrete, concrete containing carbon fibers and silica fume reduces the driving voltage required for cathodic protection by 18% compared to plain concrete, and by 28% compared to concrete with silica fume. Because of the decrease in resistivity associated with carbon fiber addition (1.1 vol. %) to mortar, overlay (embedding titanium wires for electrical contacts to steel-reinforced concrete) in the form of mortar containing carbon fibers and latex reduces the driving voltage required for cathodic protection by 10% compared to plain mortar overlay. In spite of the low resistivity of mortar overlay with carbon fibers, cathodic protection requires multiple metal electrical contacts embedded in the mortar at a spacing of 11 cm or less.[5]

9.6 STEEL REPLACEMENT

The replacement of steel rebars by fiber-reinforced polymer rebars is an emerging technology that is attractive because of the corrosion resistance of fiber-reinforced polymer.[102-107] However, this technology suffers from high cost, the poor bonding between concrete and the fiber-reinforced polymer rebar, and the low ductility of the fiber-reinforced polymer.

9.7 CONCLUSION

Methods of corrosion control of steel-reinforced concrete include steel surface treatment, the use of admixtures in concrete, surface coating on concrete, and cathodic protection.

ACKNOWLEDGMENT

This work was supported in part by the U.S. National Science Foundation.

REFERENCES

1. B.S. Wyatt, *Corros. Sci.* 35(5-8), Pt. 2, 1601-1615 (1993).
2. S.C. Das, *Struct. Eng.* 71(22), 400-403 (1993).
3. J.S. Tinnea and R.P. Brown, Proc. 1996 4th Materials Engineering Conf., Vol. 2, pp. 1531-1539 (1996).
4. I. Solomon, M.F. Bird, and B. Phang, *Corros. Sci.* 35(5-8), Pt. 2, 1649-1660 (1993).
5. J. Hou and D.D.L. Chung, *Cem. Concr. Res.* 27(5), 649-656 (1997).
6. R.J. Kessler, R.G. Powers, and I.R. Lasa, *Mater. Perform.* 37(1), 12-15 (1998).
7. R.J. Brousseau and G.B. Pye, *ACI Mater. J.* 94(4), 306-310 (1997).
8. F. Papworth and R. Ratcliffe, *Concr. Int.* 16(10) 39-44 (1994).
9. K.E.W. Coulson, T.J. Barlo, and D.P. Werner, *Oil & Gas J.* 89(41), 80-84 (1991).
10. V. Dunlap, Proc. Conf. Cathodic Protection of Reinforced Concrete Bridge Decks, NACE, pp. 131-136 (1986).
11. B. Heuze, *Mater. Perform.* 19(5), 24-33 (1980).

12. R. Pangrazzi, W.H. Hartt, and R. Kessler, *Corrosion* 50(3) 186-196 (1994).
13. R.E. Weyers, W. Pyc, and M.M. Sprinkel, *ACI Mater. J.* 95(5), 546-557 (1998).
14. S.W. Poon and I.F. Tasker, *Asia Engineer* 26(8), 17 (1998).
15. R.D. Lampton Jr. and D. Schemberger, Materials for the New Millenium, Proc. 1996 4th Materials Engineering Conf., Vol. 2, pp. 1209-1218 (1996).
16. J.L. Smith and Y.P. Virmani, *Public Roads* 60(2), 6-12 (1996).
17. B. Neffgen, *Eur. Coat. J.* (10), 700-703 (1996).
18. J. Shubrook, *Plant Eng.* 50(10), 99-100 (1996).
19. J.S. McHattie, I.L. Perez, and J.A. Kehr, *Cem. Concr. Compos.* 18(2), 93-103 (1996).
20. J. Hartley, *Steel Times* 224(1), 23-24 (1996).
21. K. Thangavel, N.S. Rengaswamy, and K. Balakrishnan, *Indian Concr. J.* 69(5), 289-293 (1995).
22. L.K. Aggarwal, K.K. Asthana, and R. Lakhani, *Indian Concr. J.* 69(5), 269-273 (1995).
23. H.O. Hasan, J.A. Ramirez, and D.B. Cleary, *Better Roads* 65(5), (1995).
24. K. Kahhaleh, J. Jirsa, R. Carrasquillo, and H. Wheat, Proc. 3rd Materials Engineering Conf., 804, pp. 8-15 (1994).
25. R.G. Mathey and J.R. Clifton, *Proc. Struct. Congr. 94*, pp. 109-115 (1994).
26. P. Schiessl and C. Reuter, *Beton — und Stahlbetonbau* 87(7), 171-176 (1992).
27. R. Korman, *ENR* 228(19), 9 (1992).
28. K.C. Clear, *Concr. Int.* 14(5), 58, 60-62 (1992).
29. T.E. Cousins, D.W. Johnston, and P. Zia, *ACI Mater. J.* 87(4), 309-318 (1990).
30. K.W.J. Treadaway and H. Davies, *Struct. Eng.* 67(6), 99-108 (1989).
31. R.A. Treece and J.O. Jirsa, *ACI Mater. J.* 86(2), 167-174 (1989).
32. S. Muthukrishnan and S. Guruviah, *Trans. SAEST* 23(2-3), 183-188 (1988).
33. T.D. Lin, R.I. Zwiers, S.T. Shirley, and R.G. Burg, *ACI Mater. J.* 85(6), 544-550 (1988).
34. H.A. El-Sayed, F.H. Mosalamy, A.F. Galal, and B.A. Sabrah, *Corros. Prev. Control* 35(4), 87-92 (1988).
35. D.P. Gustafson, *Civ. Eng.* 58(10), 38-41 (1988).
36. L. Salparanta, *Valt Tek Tutkimuskeskus Tutkimuksia* 521, 48 p (1988).
37. R.J. Higgins, *Concr. Plant Prod.* 5(6), 197-198 (1987).
38. R.J. Higgins, *Concr. Plant Prod.* 5(4), 131-132 (1987).
39. H.A. El-Sayed, M.M. Kamal, S.N. El-Ebiary, and H. Shahin, *Corros. Prev. Control* 34(1), 18-23 (1987).
40. B.W. McLean and A.J.R. Bridges, *U.K. Corrosion '85*, Vol. 2, pp. 11-17 (1985).
41. S.L. Lopata, *ASTM Special Technical Publication 841*, pp. 5-9 (1983).
42. T. Arai, K. Shirakawa, N. Mikami, S. Koyama, and A. Yamazaki, *Sumitomo Metals* 36(3), 53-71 (1984).
43. J. Clifton, CIB 83, 9th CIB Congress, pp. 68-69 (1983).
44. K. Kobayashi and K. Takewaka, *Int. J. Cem. Compos. Lightweight Concr.* 6(2), 99-116 (1984).
45. V.R. Subramanian, *Indian Concr. J.* 70(7), 383-385 (1996).
46. S.R. Yeomans, *Hong Kong Inst. Eng. Trans.* 2(2), 17-28 (1995).
47. N. Gowripalan and H.M. Mohamed, *Cem. Concr. Res.* 28(8), 1119-1131 (1998).
48. O.A. Kayyali and S.R. Yeomans, *Constr. Build. Mater.* 9(4), 219-226 (1995).
49. S.R. Yeomans, *Corrosion* 50(1), 72-81 (1994).
50. F.H. Rasheeduzzafar, F.H. Dakhil, M.A. Bader, and M.M. Khan, *ACI Mater. J.* 89(5), 439-448 (1992).

51. N.S. Rengaswamy, S. Srinivasan, and T.M. Balasubramanian, *Trans. SAEST* 23(2-3), 163-173 (1988).
52. D.B. McDonald, Y.P. Virmani, and D.F. Pfeifer, *Concr. Int.* 18(11), 39-43 (1996).
53. H. Kishikawa, H. Miyuki, S. Hara, M. Kamiya, and M. Yamashita, *Sumitomo Search* (60), 20-26 (1998).
54. J. Hou, X. Fu, and D.D.L. Chung, *Cem. Concr. Res.* 27(5), 679-684 (1997).
55. N.S. Rengaswamy, R. Vedalakshmi, and K. Balakrishnan, *Corros. Prev. Control* 42(6), 145-150 (1995).
56. J.M. Gaidis and A.M. Rosenberg, *Cem. Concr. Aggregates* 9(1), 30-33 (1987).
57. N.S. Berke, *Concr. Int.* 13(7), 24-27 (1991).
58. C.K. Nmai, S.A. Farrington, and G.S. Bobrowski, *Concr. Int.* 14(4), 45-51 (1992).
59. N.S. Rengaswamy, V. Saraswathy, and K. Balakrishnan, *J. Ferrocement* 22(4), 359-371 (1992).
60. C.K. Nmai, M.A. Bury, and H. Farzam, *Concr. Int.* 16(4), 22-25 (1994).
61. C.K. Nmai, M.A. Bury, and H. Farzam, *Concr. Int.* 16(4), 22-25 (1994).
62. S.H. Tantawi, *J. Mater. Sci. Technol.* 12(2), 95-99 (1996).
63. R.J. Scancella, Proc. 4th Materials Engineering Conf., Vol. 2, pp. 1276-1280 (1996).
64. J.R. Miller and D.J. Fielding, *Concr. Int.* 19(4), 29-34 (1997).
65. I.Z. Selim, *J. Mater. Sci. Technol.* 14(4), 339-343 (1998).
66. D. Bjegovic and B. Miksic, *Mater. Perform.* 38(11), 52-56 (1999).
67. M. Tullmin, L. Mammoliti, R. Sohdi, C.M. Hansson, and B.B. Hope, *Cem. Concr. Aggregates* 17(2), 134-144 (1995).
68. C.M. Hansson, L. Mammoliti, and B.B. Hope, *Cem. Concr. Res.* 28(12), 1775-1781 (1998).
69. N.S. Berke, M.P. Dallaire, R.E. Weyers, M. Henry, J.E. Peterson, and B. Prowell, ASTM Special Technical Publication, No. 1137, pp. 300-327 (1991).
70. N.S. Berke and A. Rosenberg, *Transp. Res. Rec.* (1211), 18-27 (1989).
71. N.S. Berke, *Transp. Res. Rec.* (1204), 21-26 (1988).
72. J. Hou and D.D.L. Chung, *Corros. Sci.* 42(9), 1489-1507 (2000).
73. J.G. Cabrera and P.A. Claisse, *Constr. Build. Mater.* 13(7), 405-414 (1999).
74. O.E. Gjorv, *ACI Mater. J.* 92(6), 591-598 (1995).
75. J.G. Cabrera, P.A. Claisse, and D.N. Hunt, *Constr. Build. Mater.* 9(2), 105-113 (1995).
76. N.R. Jarrah, O.S.B. Al-Amoudi, M. Maslehuddin, O.A. Ashiru, and A.I. Al-Mana, *Constr. Build. Mater.* 9(2), 97-103 (1995).
77. T. Lorentz and C. French, *ACI Mater. J.* 92(2), 181-190 (1995).
78. S.A. Khedr and A.F. Idriss, *J. Mater. Civ. Eng.* 7(2), 102-107 (1995).
79. C. Ozyildirim, *ACI Mater. J.* 91(2), 197-202 (1994).
80. J.T. Wolsiefer, Sr., Proc. ASCE National Convention and Exposition, pp. 15-29 (1993).
81. K. Torii, M. Kawamura, T. Asano, and M. Mihara, *Zairyo/J. Soc. Mater. Sci., Japan* 40(456), 1164-1170 (1991).
82. H.T. Cao and V. Sirivivantnanon, *Cem. Concr. Res.* 21(2-3), 316-324 (1991).
83. Anon., *Concr. Constr.* 33(2), 6 (1988).
84. K.P. Fischer, O. Bryhn, and P. Aagaard, *Publikasjon — Norges Geotekniske Institutt* (161), (1986).
85. O. Gautefall and O. Vennesland, *Nord. Concr. Res.* (2), 17-28 (1983).
86. P.J.M. Monteiro, O.E. Gjorv, and P.K. Mehta, *Cem. Concr. Res.* 15(5), 781-784 (1985).
87. O. Vennesland and O.E. Gjorv, Publication SP — American Concrete Institute 79, Vol. 2, pp. 719-729 (1983).

88. A.H. Ali, *Corros. Prev. Control* 46(3), 76-81 (1999).

89. A.H. Ali, B. El-Sabbagh, and H.M. Hassan, *Corros. Prev. Control* 45(6), 173-180 (1998).

90. M. Maslehuddin, H. Saricimen, and A.I. Al-Mana, *ACI Mater. J.* 84(1), 42-50 (1987).

91. M. Maslehuddin, A.I. Al-Mana, M. Shamim, and H. Saricimen, *ACI Mater. J.* 86(1), 58-62 (1989).

92. O.S.B. Al-Amoudi, M. Maslehuddin, and I.M. Asi, *Cem. Concr. Aggregates* 18(2), 71-77 (1996).

93. M. Maslehuddin, A.I. Al-Mana, H. Saricimen, and M. Shamim, *Cem. Concr. Aggregates* 12(1), 24-31 (1990).

94. K. Babaei and N.M. Hawkins, ASTM Special Technical Publication, No. 1137, pp. 140-154 (1991).

95. S.H. Okba, A.S. El-Dieb, and M.M. Reda, *Cem. Concr. Res.* 27(6), 861-868 (1997).

96. S.X. Wang, W.W. Lin, S.A. Ceng, and J.Q. Zhang, *Cem. Concr. Res.* 28(5), 649-653 (1998).

97. F. Andrews-Phaedonos, Transport Proc. 19th ARRB Conference of the Australian Road Research Board, pp. 245-262 (1998).

98. J.-Z. Zhang, I.M. McLoughlin, and N.R. Buenfeld, *Cem. Concr. Compos.* 20(4), 253-261 (1998).

99. R.N. Swamy and S. Tanikawa, *Mater. Struct.* 26(162), 465-478 (1993).

100. B.G. Callaghan, *Corros. Sci.* 35(5-8), Pt. 2, 1535-1541 (1993).

101. K. Kumar, R. Vedalakshmi, S. Pitchumani, A. Madhavamayandi, and N.S. Rengaswamy, *Indian Concr. J.* 70(7), 359-364 (1996).

102. K. Murphy, S. Zhang, and V.M. Karbhari, International SAMPE Symposium & Exhibition, 44(II), 2222-2230 (1999).

103. H. Saadatmanesh and F. Tannous, Proc. International Seminar on Repair and Rehabilitation of Reinforced Concrete Structures: The State of the Art, pp. 120-133 (1998).

104. S.S. Faza, Materials for the New Millennium, Proc. 4th Materials Engineering Conf., Vol. 2, pp. 905-913 (1996).

105. S. Loud, *SAMPE J.* 32(1), (1996).

106. T.R. Gentry and M. Husain, *J. Compos. Constr.* 3(2), 82-86 (1999).

107. H.C. Boyle and V.M. Karbhari, *Polym.-Plast. Technol. Eng.* 34(5), 697-720 (1995).

10 Applications of Submicron-Diameter Carbon Filaments

CONTENTS

SYNOPSIS The applications of submicron-diameter carbon filaments grown catalytically from carbonaceous gases are reviewed. The applications relate to structural applications, EMI shielding, electromagnetic reflection, surface electrical conduction, DC electrical conduction, field emission, electrochemical applications, thermal conduction, strain sensors, porous carbons, and catalyst support.

RELEVANT APPENDIX: *J*

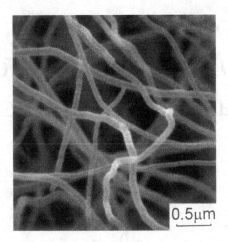

FIGURE 10.1 Scanning electron micrograph of carbon filaments.

10.1 INTRODUCTION

Submicron-diameter carbon filaments are grown catalytically from carbonaceous gases at 500–700°C,[1,2] although they include the carbon nanotubes,[3] which have diameters in the nanometer range. Due to the higher yield in production, the former are more abundant, and applications are more well developed, than the latter. Submicron carbon filaments are to be distinguished from conventional carbon fibers, which are made by pyrolysis of pitch or polymer.[4-6] They are also to be distinguished from vapor-grown carbon fibers (VGCF), which are prepared by pyrolysis of carbonaceous gases to noncatalytically deposit carbon on catalytically grown submicron-diameter carbon filaments at 950–1100°C.[7-14] Carbon filaments differ from both conventional and vapor-grown carbon fibers in their small diameters. Conventional carbon fibers typically have diameters around 10 μm, and VGCF have diameters up to 10 μm. Both carbon filaments and VGCF are not continuous, though the latter can be longer than the former. In contrast, conventional carbon fibers can be continuous. In spite of the discontinuous nature of carbon filaments, the aspect ratio can be quite high because of the small diameter. Carbon filaments tend not to be straight. Thus, they are intertwined and have a morphology that resembles cotton wool (Figure 10.1). Moreover, they are typically disordered crystallographically, unless they have been graphitized by heat treatment. However, even a disordered filament exhibits a fishbone morphology for the carbon layers (Figure 10.2). Each filament has an axial hollow channel in the middle due to the catalyst particle used in filament growth. In this chapter, the term "filaments" refers to filaments of submicron diameter, whereas "fibers" refers to fibers of diameter greater than 1 μm.

Carbon filaments are commercially available, though not in large volumes. As the amount of usage increases, their price will fall. According to Applied Sciences, Inc., which manufactures catalytically grown carbon filaments, the price will fall to U.S. $2-3 per pound. This is even lower than that of short isotropic pitch-based carbon fibers. In order for usage to increase, applications must be developed.

Channel

FIGURE 10.2 Schematic of a carbon filament showing the fishbone morphology of the carbon layers and the hollow channel along the axis of the filament.

Much research has been conducted to understand the process of catalytic growth of carbon filaments.[1,2] However, relatively little attention has been given to the applications of submicron-diameter carbon filaments, although considerable progress has been made recently.

10.2 STRUCTURAL APPLICATIONS

Structural applications are most important for conventional continuous carbon fibers. Due to their discontinuous nature, carbon filaments are far less effective than continuous carbon fibers as a reinforcement in composites. Because of their small diameter and consequent large area of the interface between filaments and matrix in a composite, and because of their texture with the carbon layers at an angle from the filament axis, carbon filaments are not as effective as a reinforcement as short carbon fibers at the same volume fraction, as shown for both thermoplast[15,16] and cement matrices.[17] For example, in a thermoplast matrix, carbon filaments at 19 vol.% give a tensile strength of 27 MPa,[16] whereas carbon fibers (isotropic pitch-based, 3000 μm long) at 20 vol.% give a tensile strength of 64 MPa.[15] In a cement paste matrix, carbon filaments at 0.51 vol.% give a tensile strength of 1.2 MPa, whereas carbon fibers (isotropic pitch-based, 5 mm long) give a tensile strength of 1.7 MPa.[17] Although the filaments do not reinforce as well as the fibers, they still reinforce under tension. In a cement matrix, the tensile strength, modulus and ductility, and the compressive modulus are all increased by the filaments (0.51 vol.%), but the compressive strength and ductility are decreased by the filaments.[17] More success has been found with hybrid composites involving both carbon filaments and conventional carbon fibers[18] than in composites involving carbon filaments only.[16,17] The use of carbon filaments in the interlaminar region between adjacent layers of conventional continuous carbon fibers in an epoxy-matrix composite enhances both transverse and longitudinal vibration damping ability, and increases the storage modulus in the transverse direction. Even though the filament volume fraction (0.6%) is negligibly low compared to the fiber volume fraction (56.5%) in the hybrid composite, the filaments are effective. The longitudinal storage modulus is only slightly decreased by the filament addition.[18] Another form of hybrid composite involves the catalytic growth of carbon filaments on conventional carbon fibers to provide mechanical interlocking between adjacent fibers in a composite.[19,20] The interfacial shear strength between fiber and a polymer matrix is increased by over 4.75 times after the growth of filaments on the fiber.[20] Furthermore, the specific surface area is increased from about 1.0 m^2/g to 250–300 m^2/g.[20] However, the practicality of composite fabrication using such hybrid fibers remains an issue.

Due to the clingy morphology and small diameter of carbon filaments, dispersion of the filaments in a composite requires more care than that of conventional short carbon fibers. A method of dispersing the carbon filaments in a thermoplast matrix involves (1) dispersing the filaments in an isopropyl alcohol aqueous solution with the help of a trace amount of a dispersant, such as Triton X102® from Rohm and Haas Co.; (2) mixing the slurry with thermoplast powder at room temperature by using a kitchen blender such that the concentration of isopropyl alcohol in the aqueous solution is adjusted so that the thermoplast particles are suspended; (3) draining the solution; (4) drying at 120°C; and (5) hot-pressing uniaxially above the glass transition temperature of the thermoplast and at 1000 psi (6.9 MPa) for 0.5 h. The mixing in Step 2 causes very little filament breakage, so the aspect ratio of the filaments after mixing remains high (typically > 1000).[15,20] In this method, the thermoplast must be in the form of fine particles because of the small diameter of the carbon filaments. As thermoplasts are in the form of pellets rather than particles, the choice is limited. In the case of a thermosetting resin such as epoxy, dispersion of the carbon filaments requires dilution of the resin with a solvent to lower the viscosity, and subsequent mixing of the filament-resin slurry by using a vigorous means, such as a blender. Due to the strong effect of the form of the matrix raw material on the dispersion of the carbon filaments, the properties (both mechanical and electromagnetic) of the composites depend significantly on the matrix material.

Catalytically grown carbon filaments have a layer of polyaromatic hydrocarbons on their surfaces, because of the process in which they are grown.[22] The hydrocarbon layer can be removed by cleansing with a solvent, such as acetone or methylene chloride.[22] The removal or partial removal of the hydrocarbon layer improves the bonding between carbon filaments and a thermoplast matrix, as suggested by the fact that the volume electrical resistivity of the composite is much lower when the hydrocarbon layer has been removed before incorporating the filaments in the composite,[21,23] and as supported by the mechanical properties of the composites.[24]

Surface treatment of the filaments by oxidation helps the mechanical properties of cement-matrix and polymer-matrix composites.[25,26] The bond between carbon and a cement matrix is weak compared to that between carbon and a polymer matrix. Therefore, surface treatment of carbon for improving the bond with cement is particularly important. The treatment of catalytically grown carbon filaments (0.1 μm diameter) with ozone gas (0.3 vol.% in air, 160°C, 10 min) increases the tensile strength, modulus and ductility, and the compressive strength and modulus and ductility of cement pastes relative to the values for pastes with the same volume fraction of untreated filaments.[25] Similar effects apply to the ozone treatment of carbon fibers,[26,27] for which it has been shown that ozone treatment improves wetting by water, degree of fiber dispersion in cement, and bond strength with cement, in addition to increasing surface oxygen concentration.[27]

Ceramic-matrix (Al_2O_3 or $MgAl_2O_4$ as matrix) composite powders that contain *in situ* formed carbon nanotubes may be useful for structural applications.[28]

10.3 ELECTROMAGNETIC INTERFERENCE SHIELDING, ELECTROMAGNETIC REFLECTION, AND SURFACE ELECTRICAL CONDUCTION

EMI shielding[29-32] is in critical demand due to the interference of wireless devices with digital devices, and the increasing sensitivity and importance of electronic devices. EMI shielding is one of the main applications of conventional short carbon fibers.[33] Because of the small diameter, carbon filaments are more effective for EMI shielding at the same volume fraction in a composite than conventional short carbon fibers, as shown for both thermoplast[15,16] and cement[17,34] matrices. For example, in a thermoplast matrix, carbon filaments at 19 vol.% give an EMI shielding effectiveness of 74 dB at 1 GHz,[16] whereas carbon fibers at 20 vol.% give a shielding effectiveness of 46 dB at 1 GHz.[15] In a cement-matrix composite, fiber volume fractions are typically less than 1%. Carbon filaments at 0.54 vol.% in a cement paste give an effectiveness of 26 dB at 1.5 GHz,[17] whereas carbon fibers at 0.84 vol.% in a mortar yield an effectiveness of 15 dB at 1.5 GHz.[34] These effectiveness measurements were made with the same fixture and about the same sample thicknesses. A low-volume fraction of the filler is attractive for maintaining ductility or resilience in the polymer-matrix composite, as both ductility and resilience decrease with increasing filler volume fraction. Resilience is particularly important for EMI-shielding gaskets and electric cable jackets. In addition, a low-volume fraction of the filler reduces the material cost and improves the processability of the composites, whether polymer-matrix or cement-matrix.

The greater shielding effectiveness of filaments compared to fibers is because of the skin effect, i.e., high-frequency electromagnetic radiation interacts only with the near surface region of an electrical conductor. However, carbon filaments are still not as effective as nickel fibers of 2 μm diameter at the same volume fraction, as shown for a thermoplast matrix.[23] On the other hand, by coating a carbon filament with nickel by electroplating, a nickel filament (0.4 μm diameter) with a carbon core (0.1 μm diameter) is obtained.[23,35] Nickel filaments are more effective than nickel fibers for shielding, because of their small diameters. At 1 GHz, a shielding effectiveness of 87 dB was attained by using only 7 vol.% nickel filaments in a thermoplast matrix.[23] The shielding is almost all by reflection rather than absorption.

The high radio wave reflectivity of carbon filament-reinforced cement paste makes carbon filament concrete attractive for use for lateral guidance in automatic highways.[36] Automatic highways are those that provide fully automated control of vehicles so that safety and mobility are enhanced. In other words, a driver does not need to drive on an automatic highway; the vehicle goes automatically, with both lateral control (steering to control position relative to the center of the traffic lane) and longitudinal control (speed and headway). Current technology uses magnetic sensors with magnetic highway markings to provide lateral guidance, and uses radar to monitor vehicle position relative to other vehicles in its lane for the purpose of longitudinal guidance. Cement paste containing 0.5 vol.% carbon filaments exhibits reflectivity at 1 GHz that is 29 dB higher than the transmissivity. Without the filaments, the reflectivity is 3–11 dB lower than the transmissivity.

Compared to magnetic technology, the attractions of electromagnetic technology are low material cost, low labor cost, low peripheral electronic cost, good mechanical properties, good reliability, and high durability. Moreover, the magnetic field from a magnetic marking can be shielded by electrical conductors between the marking and the vehicle, whereas the electromagnetic field cannot be shielded easily.

The surface impedance of carbon filament composites, nickel filament composites, and nickel fiber composites is low. In particular, at 1 GHz, the surface impedance is comparable to that of copper for a thermoplast-matrix composite with 7 vol.% nickel filaments and a thermoplast-matrix composite with 13 vol.% nickel fibers.[16] The surface impedance is higher for carbon filament composites than nickel filament composites or nickel fiber composites at similar filler volume fractions.[16] Although carbon filaments have a lower density than nickel filaments, a thermoplast-matrix composite with 7 vol.% nickel filaments has the same specific surface conductance as one with 19 vol.% carbon filaments.[16] The low surface impedance is valuable for applications related to electrostatic discharge protection and microwave waveguides.

10.4   DC ELECTRICAL CONDUCTION

Submicron-diameter carbon filaments are useful as an electrically conducting additive for enhancing the DC electrical conductivity of a polymer-matrix composite,[37-41] provided they are properly dispersed.[16,21,39] An application that benefits from the enhanced conductivity relates to solid rocket propellants, for which enhanced conductivity decreases the incidence of dangerous electric discharge events.[38] Another application pertains to molecular optoelectronics and involves the use of carbon nanotubes.[40,41]

10.5   FIELD EMISSION

The high aspect ratio, small radius of curvature at the tips, high chemical stability, and high mechanical strength of carbon nanotubes are advantageous for field emission, i.e., the emission of electrons under an applied electric field. Field emission is relevant to various electronic devices, including high-current electron sources, flat-panel displays, and light source bulbs.[42-53] Nanotubes as emitters provide significantly brighter displays than either cathode ray tubes or Spindt tip-based displays.[3] In addition, field emission electron sources are energy saving compared with thermionic ones, because no heating is necessary to emit electrons from the cathode surface.[45] Moreover, carbon nanotubes are free of any precious or hazardous element.[45] Alignment of the nanotubes is desired for this application, and they should have closed, well-ordered tips.

10.6 ELECTROCHEMICAL APPLICATION

Due to its electrical conductivity and chemical resistance, carbon is an important material for electrochemical applications, particularly electrodes for electrochemical cells,[54-62] double-layer capacitors,[61,63-67] and energy storage.[68,69] The small diameter of carbon nanotubes is advantageous for microelectrode arrays.[70-73] Capacitors exhibiting fast response (100 Hz) and high specific capacitance (100 F/g) have been attained by using carbon filaments.[68] Porous tablets of carbon nanotubes have been fabricated by using polymeric binders as polarizable electrodes in capacitors.[67]

Electrolyte absorptivity, specific surface area, surface chemistry, and crystallographic structure are important for electrodes. Carbon black[59] is the most common type of carbon for these applications, though the use of conventional carbon fibers,[54-60,63-66] VGCF,[61,62] and carbon filaments[68,74-77] has been investigated. The removal of the hydrocarbon layer on carbon filaments improves electrochemical behavior, as indicated by the electron transfer rate across the electrode-electrolyte interface.[22]

Catalytically grown carbon filaments of diameter 0.1 μm have been shown superior to carbon black in lithium primary cells that use carbon as a porous electrode (current collector),[75] and as an electrically conductive additive in a nonconducting electrode.[76] The current collector of the lithium/thionyl chloride (Li/SOCl$_2$) cell conventionally uses carbon black, which needs a teflon binder. Because of the cleansing ability of thionyl chloride, carbon filaments used in place of carbon black do not require solvent cleansing before use. As the filaments tend to cling together, a binder is not necessary, in contrast to carbon black. Using the same paper-making process, carbon filaments can be made into a thinner sheet than carbon black. The thinness is valuable for enhancing the energy density of the cell, as the area over which the lithium anode faces the carbon current collector is increased. In addition, the packing density is lower for the filament sheet than the carbon black sheet, so the catholyte absorptivity is higher for the filament sheet than the carbon black and, consequently, energy density is further increased.[75]

The MnO$_2$ cathode of a Li/MnO$_2$ primary cell is electrically nonconducting, so a conductive additive, typically carbon black, is mixed with the MnO$_2$ particles. The use of catalytically grown carbon filaments of diameter 0.1 μm in place of carbon black causes the running voltage near cell end-of-life to decline gradually, in contrast to the abrupt end-of-life when carbon black is used. The gradualness toward end-of-life is due to a high electron transfer rate and a high rate of electrolyte absorption. For the filaments to be effective, they need to undergo solvent cleansing prior to use.[76]

By using catalytically grown carbon filaments with diameters around 80 Å, a double-layer capacitor of specific capacitance 102 F/g at 1 Hz has been achieved.[77]

The catalytic growth of carbon filaments on carbons provides a way of modifying the surface of carbons to improve electrochemical behavior.[78] The resulting carbons are called "hairy carbons." Particularly abundant hair growth occurs when the carbon is carbon black because of the confinement of the catalyst size by the pores in the carbon black. Hair growth, followed by an oxidation heat treatment, results in further improvement in electrochemical behavior. The particulate nature of hairy carbon black is in contrast to the fibrous nature of carbon filaments or carbon fibers. The particulate nature facilitates dispersion, while the hairiness makes a binder unnecessary for electrode forming.

10.7 THERMAL CONDUCTION

Due to the low temperature (500–700°C) during the catalytic growth of carbon filaments, they may be only slightly crystalline (i.e., almost totally amorphous) after fabrication. However, subsequent heat treatment at 2500–3000°C causes graphitization,[22] which is expected to result in a large increase in thermal conductivity. A high thermal conductivity is valuable for use of the filaments in composites for thermal management, which is critically needed for heat dissipation from electronic packages, space radiators, and plasma facing. Because of the small diameters of the carbon filaments, single-filament thermal conductivity measurement is difficult. The thermal conductivity of conventional pitch-based carbon fiber is 603 W/m.K for P-120,[79] 750 W/m.K for P-100-4,[80] 1000 W/m.K for P-X-5,[80] and 1055 W/m.K for K1100;[79] that of VGCF is 2540–2680 W/m.K.[80]

The thermal conductivity of carbon filament composites has not been reported, but polymer-matrix,[81] aluminum-matrix,[82,83] and carbon-matrix[82,84-86] composites containing VGCF exhibit thermal conductivities up to 466, 642, and 910 W/m.K, respectively. A polymer-matrix composite containing conventional pitch-based continuous carbon fibers (P-120) exhibits thermal conductivity 245 W/m.K.[80]

Although the thermal conductivity has not been reported for carbon filament composites, the volume electrical resistivity has, and a low electrical resistivity is expected to correlate with a high thermal conductivity. The electrical resistivity of a carbon filament polymer-matrix composite is higher than that of a nickel filament composite or a nickel fiber composite at the same filler volume fraction and with the same matrix polymer.[23] At 13 vol.% carbon filaments, the DC resistivity of the composite is 0.37 Ω.cm; at 13 vol.% nickel filaments, the resistivity is 0.0035 Ω.cm.[23] The high resistivity of the carbon filament composite is attributed to the high resistivity of the carbon filaments, which have not been graphitized, and to the large filament-matrix interface area per unit volume due to the small diameters of the carbon filaments (0.1 μm) compared to the nickel filaments (0.4 μm diameter).

10.8 STRAIN SENSORS

Because of the advent of smart structures, strain sensors are increasingly needed for structural vibration control and *in situ* structural health monitoring. Composites containing conventional short carbon fibers have their volume electrical resistivity change reversibly upon reversible strain, allowing them to serve as strain sensors. In the case of a composite with a ductile matrix (such as a polymer matrix), this phenomenon is due to the change in the distance between adjacent fibers in the composite and is referred to as piezoresistivity.[87] Tension causes this distance to increase, thereby increasing resistivity; compression causes this distance to decrease, decreasing resistivity. In the case of a composite with an elastomer matrix, the phenomenon is different in both direction and origin; the resistivity decreases upon tension, as observed for a silicone-matrix composite with 0.4 µm diameter nickel filaments (with a 0.1 µm diameter carbon filament core in each nickel filament[23]).[88] This reverse piezoresistivity effect is probably due to the increase in filament alignment upon tension. In the case of a composite with a brittle matrix, the phenomenon is not reverse but is different in origin; it is due to the slight (< 1 µm) pullout of the fiber bridging a crack as the crack opens, and the consequent increase in the contact electrical resistivity of the fiber-matrix interface.[89-91] Tension causes a crack to open, increasing the resistivity; compression causes a crack to close, decreasing the resistivity.

The use of carbon filaments in place of conventional short carbon fibers in a polymer-matrix composite improves the reproducibility and linearity of the piezoresistivity effect.[92] This is because of the small diameter of the filaments, which results in (1) a large number of filaments per unit volume of the composite, (2) reduced tendency of the filaments to buckle upon compression of the composite, and (3) reduced tendency for the matrix at the junction of adjacent filaments to be damaged. Furthermore, the use of the filaments enhances the tendency toward the reverse piezoresistivity effect.[93]

The use of carbon filaments in place of conventional short carbon fibers in a cement-matrix composite results in increased noise in the electromechanical effect.[17] This is because of the bent morphology and large aspect ratio of the filaments, which hinder the pullout of filaments. Thus, carbon filaments are not attractive for cement-matrix composite strain sensors.

10.9 POROUS CARBONS

Porous carbons with high porosity (above 50 vol.%) and/or high specific surface area are useful for numerous nonstructural applications such as electrodes, catalysts, catalyst support, filters, chemical absorbers, molecular sieves, membranes, dental and surgical prosthetic devices, and thermal insulators.[94-102] They can be made from carbon fibers, which may be bound by a binder such as a polymer, pitch, or carbon. Alternatively, they can be made by carbonizing organic fibers that are bound by a binder. In either case, a pore-forming agent may be used, although it is not essential. Porous carbons can also be made from a polymer (such as a phenolic) that is not in the form of fibers, through foaming and carbonization. Due to the large diameters (typically 10 µm or more) of the carbon or organic fibers for fiber-based porous carbons, and the foaming process for polymer-based porous carbons, the pores in the resulting porous carbons are large (> 40 µm in mean size). As a result, porous carbons are low in strength (< 7 MPa under compression) and in specific geometric surface area (SGSA, < 1100 cm²/cm³). This is more serious for polymer-based porous carbons than for fiber-based porous carbons. By using carbon filaments in place of carbon fibers, a porous carbon of mean pore size 4 µm, SGSA 35,000 cm²/cm³, and compressive strength 30–35 MPa has been obtained.[103]

Carbons with high specific surface area include the conventional activated carbon bulk,[104-106] activated carbon fibers,[107] fine carbon particles,[108] carbon aerogels,[109] and carbon nanotubes.[110] Other than adsorbents for purification and chemical processing,[104-106] these carbons are used as catalytic materials, battery electrode materials, capacitor materials, gas (e.g., hydrogen) storage materials, and biomedical engineering materials. A hydrogen storage capacity of approximately 1.5 wt. % has been reached at ambient temperature and a hydrogen pressure of 125 bar.[110]

A problem concerning porous carbon materials relates to the need for high-surface-area porous carbon materials with mesopores and/or macropores for some applications; many macromolecules and ions encountered in purification, catalysis, and batteries cannot penetrate the surface of the carbon without such pores.

According to IUPAC, pores are classified into four types, namely macropores (diameter > 500 Å), mesopores (20 Å $<$ diameter < 500 Å), micropores (8 Å $<$ diameter < 20 Å), and micro-micropores (diameter < 8 Å). Most pores are micropores in conventional activated carbons. The pore volume in activated carbon fibers (including pitch-based, PAN-based, and rayon-based) is occupied by micropores (mainly) and micro-micropores. The pore volume in carbon aerogels is occupied by mesopores (mainly) and micropores. On the other hand, the specific surface areas of carbon aerogels are low (e.g., 650 m^2/g^{106}) compared to activated carbons (as high as 3000 m^2/g).

A mesoporous carbon (83% of total pore volume being > 30 Å pore size, 17% of total pore volume being < 30 Å pore size) that has a total pore volume of 1.55 cm^3/g and high specific surface area (1310 m^2/g) is in the form of carbon filaments that have been surface-oxidized in ozone at 150°C and then activated in $CO_2 + N_2$ (1:1) at 970°C.[111] Without activation, the filaments have only 44–57% of the total pore volume being > 30 Å pore size, and the specific surface area is low (41–54 m^2/g). Activation by CO_2 greatly increases the specific surface area. This is in contrast to conventional carbon fibers, which have essentially no pores. The porous nature of the filaments is attributed to the fact that the filaments are made from carbonaceous gases. The separation between adjacent filaments in a filament compact is of the order of 0.1 μm, thus providing macropores within the compact. These elongated macropores serve as channels that facilitate fluid flow. The combination of mesopores within each filament and macropores between the filaments is in contrast to carbon aerogels, which have micropores within each particle and between particles, and mesopores between chains of interconnected particles. The mesoporous activated carbon filaments have mean mesopore size (BJH) of 54 Å.

10.10 CATALYST SUPPORT

Catalytically grown carbon filaments, even without activation, have been shown to be effective catalyst (e.g., Ni- and Fe-based particles) support material.[112-117] The catalytic activity for the conversion of hydrocarbons is higher than that when the catalyst particles are supported on either active carbon or γ-alumina.[113,115] The dispersion of a catalyst on carbon filaments is improved by prior surface treatment of the filaments by oxidation (using nitric acid).[118] A platinum catalyst supported by carbon filaments is more active and filters more easily than that supported by activated carbon, as shown for the hydrogenation of nitrobenzene.[118]

10.11 CONCLUSION

Applications of submicron-diameter carbon filaments include structural applications, EMI shielding, electromagnetic reflection, surface electrical conduction, DC electrical conduction, electrochemical applications, thermal conduction, strain sensors, and catalyst support. Most applications involve the incorporation of the carbon filaments in composites, most commonly polymer-matrix composites. Particularly promising applications include (1) the use of carbon filaments between layers of conventional continuous carbon fibers for improving the vibration damping ability and the storage modulus in the transverse direction; (2) coating the carbon filaments with nickel to form nickel filaments for use as a filler in polymer-matrix composites for EMI shielding, electrostatic discharge protection, and microwave waveguides; (3) the use of carbon filaments as a filler in concrete for lateral guidance in automatic highways; (4) the use of carbon filaments as a porous electrode and as an electrically conducting additive in a nonconducting electrode for lithium primary cells; (5) the use of carbon filaments for double-layer capacitors; (6) the use of carbon filaments as a filler in a polymer-matrix composite strain sensor; (7) the use of carbon nanotubes and activated carbon filaments for adsorption, hydrogen storage, and catalyst support; and (8) the use of carbon nanotubes for field emission.

ACKNOWLEDGMENT

This work was supported by the Defense Advanced Research Projects Agency of the U.S. Department of Defense.

REFERENCES

1. N.M. Redriguez, *J. Mater. Res.* 8(12), 3233-3250 (1993).
2. P. Chitrapu, C.R.F. Lund and J.A. Tsamopoulos, *Carbon* 30(2), 285-293 (1992).
3. S. Subramoney, *The Electrochemical Society Interface* 18(4), 34-37 (1999).
4. D.D.L. Chung, *Carbon Fiber Composites,* Butterworth-Heinemann, Boston (1994).
5. L.H. Peebles, *Carbon Fibers,* CRC Press, Boca Raton (1994).
6. M.S. Dresselhaus, G. Dresselhaus, K. Sugihara, I.L. Spain and H.A. Goldberg, *Graphite Fibers and Filaments*, Vol. 1, Springer-Verlag, Berlin (1988).
7. M. Endo, *CHEMTECH* 18(9), 568-576 (1988).
8. G.G. Tibbetts, *Carbon* 27(5) 745-747 (1989).
9. G.G. Tibbetts, D.W. Gorkiewicz and R.L. Alig, *Carbon* 31(5), 809-814 (1993).
10. T. Kato, T. Matsumoto, T. Saito, J.-H. Hayashi, K. Kusakabe and S. Morooka, *Carbon* 31(6), 937-940 (1993).
11. S.R. Mukai, T. Masuda, T. Harada and K. Hashimoto, *Carbon* 34(5), 645-648 (1996).
12. T. Masuda, S.R. Mukai, H. Fujikawa, Y. Fujikata and K. Hashimoto, *Mater. Manuf. Proc.* 9(2), 237-247 (1994).
13. M. Ishioka, T. Okada and K. Matsubara, *J. Mater. Res.* 7(11), 3019-3022 (1992).
14. G.G. Tibbetts, SAE Transactions 99, Sect. 1, pp. 246-249 (1990).
15. L. Li and D.D.L. Chung, *Composites* 25(3), 215-224 (1994).

16. X. Shui and D.D.L. Chung, *J. Electron. Mater.* 26(8), 928-934 (1997).

17. X. Fu and D.D.L. Chung, *Cem. Concr. Res.* 26(10), 1467-1472 (1996); 27(2), 314 (1997)

18. S.W. Hudnut and D.D.L. Chung, *Carbon* 33(11), 1627-1631 (1995).

19. W.B. Downs and R.T.K. Baker, *Carbon* 29(8), 1173-1179 (1991).

20. W.B. Downs and R.T.K. Baker, *J. Mater. Res.* 10(3), 625-633 (1995).

21. X. Shui and D.D.L. Chung, 38th Int. SAMPE Symp. Exhib., Book 2, Vince Bailey, Gerald C. Janicki and Thomas Haulik, Eds., pp. 1869-1875 (1993).

22. X. Shui, C.A. Frysz and D.D.L. Chung, *Carbon* 33(12), 1681-1698 (1995).

23. X. Shui and D.D.L. Chung, *J. Electron. Mater.* 24(2), 107-113 (1995).

24. G.G. Tibbetts and J.J. McHugh, *J. Mater. Res.* 14(7), 2871-2880 (1999).

25. X. Fu and D.D.L. Chung, *Carbon* 36(4), 459-462 (1998).

26. G. Caldeira, J.M. Maia, O.S. Carneiro, J.A. Covas and C.A. Bernardo, *ANTEC '97,* 2, 2352-2356 (1997).

27. X. Fu, W. Lu and D.D.L. Chung, *Carbon* 36(9), 1337-1345 (1998).

28. Ch. Laurent, A. Peigney, O. Quenard and A. Rousset, *Key Eng. Mater.* 132-136 (Pt. 1), 157-160 (1997).

29. B.D. Mottahed and S. Manoocheheri, *Polym.-Plast. Technol. Eng.* 34(2), 271-346 (1995).

30. P.S. Neelakanta and K. Subramaniam, *Adv. Mater. Proc.* 141(3), 20-25 (1992).

31. G. Lu, X. Li and H. Jiang, *Compos. Sci. Tech.* 56, 193-200 (1996).

32. Kaynak, A. Polat and U. Yilmazer, *Mater. Res. Bull.* 31(10), 1195-1206 (1996).

33. P.B. Jana and A.K. Mallick, *J. Elastomers Plast.* 26(1), 58-73 (1994).

34. J.-M. Chiou, Q. Zheng and D.D.L. Chung, *Composites* 20(4), 379-381 (1989).

35. X. Shui and D.D.L. Chung, *J. Mater. Sci.* 35, 1773-1785 (2000).

36. X. Fu and D.D.L. Chung, *Cem. Concr. Res.* 28(6), 795-801 (1998).

37. V. Chellappa and B.Z. Jang, *J. Mater. Sci.* 30(19), 4879-4883 (1995).

38. C.W. Farriss II, F.N. Kelley and E. Von Meerwall, *J. Appl. Polym. Sci.* 55(6), 935-943 (1995).

39. J. Sandler, M.S.P. Shaffer, T. Prasse, W. Bauhofer, K. Schulte and A.H. Windle, *Polymer* 40, 5967-5971 (1999).

40. S.A. Curran, P.M. Ajayan, W.J. Blau, D.L. Carroll, J.N. Coleman, A.B. Dalton, et. al., *Adv. Mater.* 10(14), 1091-1093 (1998).

41. L. Dai, *Polymers Adv. Tech.* 10(7), 357-420 (1999).

42. R.Z. Ma, C.L. Xu, B.Q. Wei, J. Liang, D.H. Wu and D.J. Li, *Mater. Res. Bull.* 34(5), 741-747 (1999).

43. J.-M. Bonard, J.-P. Salvetat, T. Stockli, L. Forro and A. Chatelain, *Appl. Phys.* 69(3), 245-254 (1999).

44. S. Huang, L. Dai and A.W.H. Mau, *J. Phys. Chem. B.* 103(21), 4223-4227 (1999).

45. Y. Saito and S. Uemura, *Carbon* 38(2), 169-182 (2000).

46. Y. Saito, K. Hamaguchi, R. Mizushima, S. Uemura, T. Nagasako, J. Yotani, et al., *Appl. Surface Sci.* 146(1), 305-311 (1999).

47. Y. Saito, *Nihon Enerugi Gakkaishi/Journal of the Japan Institute of Energy* 77(9), 867-875 (1998) (Japanese).

48. T. Habermann, A. Goehl, K. Janischowsky, D. Nau, M. Stammler, L. Ley, et al., Proc. of the IEEE International Vacuum Microelectronics Conference, pp. 200-201 (1998).

49. K.A. Dean, P. VonAllmen and B.R. Chalamala, Proc. of the IEEE International Vacuum Microelectronics Conference, pp. 196-197 (1998).

50. Y. Saito, K. Hamaguchi, S. Uemura, K. Uchida, Y. Tasaka, F. Ikazaki, et al., *Appl. Phys.* 67(1), 95-100 (1998).

51. N.I. Sinitsyn, Y.V. Gulyaev, G.V. Torgashov, L.A. Chernozatonskii, Z.Y. Kosakovskaya, Y.F. Zakharchenko, et al., *Appl. Surface Sci.* 111, 145-150 (1997).
52. Y.V. Gulyaev, L.A. Chernozatonskii, Z.Y. Kosakovskaya, A.L. Musatov, N.I. Sinitsin and G.V. Torgashov, Proc. of the IEEE International Vacuum Microelectronics Conference, pp. 5-9 (1996)
53. N.I. Sinitsyn, Y.V. Gulyaev, N.D. Devjatkov, M.B. Golant, A.M. Alekseyenko, Y.F. Zakharchenko, et al., *Radiotekhnika* (4), 8-17 (1999) (Russian).
54. M.W. Verbrugge and D.J. Koch, *J. Electrochem. Soc.* 143(1), 24-31 (1996).
55. M. Endo, J.-I. Nakamura, A. Emori, Y. Sasabe, K. Takeuchi and M. Inagaki, Molecular Crystals and Liquid Crystals, Proc. 7th Int. Symp. on Intercalation Compounds, 244, 171-176 (1994).
56. T. Takamura, M. Kikuchi, H. Awano, U. Tatsuya, and Y. Ikezawa, Materials Research Society Symp. Proc., 393, 345-355 (1995).
57. T. Tamaki, *Ibid*, pp. 357-365.
58. R. Yazami, K. Zaghib, and M. Deschamps, *J. Power Sources* 52, 55-59 (1994).
59. O. Chusid, Y. Ein Ely, D. Aurbach, M. Babai and Y. Carmeli, *J. Power Sources* 43-44, 47-64 (1993).
60. M. Endo, Y. Nishimura, T. Takahashi, K. *J. Phys. Chem. Solids* 57(6-8), 725-728 (1996).
61. M. Endo, Y. Okada and H. Nakamura, *Synth. Met.* 34(1-3), 739-744 (1989).
62. K. Zaghib, K. Tatsumi, H. Abe, T. Ohsaki, Y. Sawad and S. Higuchi, *J. Electrochem. Soc.* 145(1), 210-215 (1998).
63. S. Biniak, B. Dzielendziak and J. Siedlewski, *Carbon* 33(9), 1255-1263 (1995).
64. Tanahashi, A. Yoshida and A. Nishino, *Carbon* 28(4), 477-482 (1990).
65. M. Ishikawa, M. Morita, M. Ihara and Y. Matsuda, *J. Electrochem. Soc.* 141(7), 1730-1734 (1994).
66. Y. Matsuda, M. Morita, M. Ishikawa and M. Ihara, *J. Electrochem. Soc.* 140(7), L109-L110 (1993).
67. R.Z. Ma, J. Liang, B.Q. Wei, B. Zhang, C.L. Xu and D.H. Wu, *J. Power Sources* 84(1), 126-129 (1999).
68. S.M. Lipka, Proc. 13th Annual Battery Conference on Applications and Advances, IEEE, pp. 373-374 (1998).
69. G. Che, B.B. Lakshmi, C.R. Martin and E.R. Fisher, *Langmuir* 15(3), 750-758 (1999).
70. M. Burghard, G. Duesberg, G. Philipp, J. Muster and S. Roth, *Adv. Mater.* 10(8), 584-588 (1998).
71. J.J. Davis, R.J. Coles and H.A.O. Hill, *J. Electroanal. Chem.* 440(1-2), 279-282 (1997).
72. N. Grobert, M. Terrones, A.J. Osborne, H. Terrones, W.K. Hsu, S. Trasobares, et al., *Appl. Phys.* 67(5), 595-598 (1998).
73. P.J. Britto, K.S.V. Santhanam, A. Rubio, J.A. Alonso and P.M. Ajayan, *Adv. Mater.* 11(2), 154-157 (1999).
74. X. Shui, D.D.L. Chung and C.A. Frysz, *J. Power Sources* 47(3), 313-320 (1994).
75. C.A. Frysz, X. Shui and D.D.L. Chung, *J. Power Sources* 58(1), 55-66 (1996).
76. C.A. Frysz, X. Shui and D.D.L. Chung, *J. Power Sources* 58(1), 41-54 (1996).
77. C. Niu, E.K. Sichel, R. Hoch, D. Moy and H. Tennent, *Appl. Phys. Lett.* 70(11), 1480-1482 (1997).
78. X. Shui, C.A. Frysz and D.D.L. Chung, *Carbon* 35(10-11), 1439-1455 (1997).
79. W.E. Lundblad, H.S. Starrett and C.W. Wanstrall, Proc. 26th Int. SAMPE Technical Conference, pp. 759-764 (1994).
80. B. Nysten and J.-P. Issi, *Composites* 21(4), 339-343 (1990).

81. J.-M. Ting, J.R. Guth and M.L. Lake, Proc. 19th Annual Conf. Composites, American Ceramic Soc., 16(4), 279-288 (1995).

82. J.-M. Ting, M.L. Lake and D.R. Duffy, *J. Mater. Res.* 10(6), 1478-1484 (1995).

83. J.-M. Ting and M.L. Lake, *J. Mater. Res.* 10(2), 247-250 (1995).

84. J.-M. Ting and M.L. Lake, *J. Nuclear Mater.* 212(1), Pt. B, 1141-1145 (1994).

85. J.-M. Ting and M.L. Lake, *Carbon* 33(5), 663-667 (1995).

86. J.-M. Ting and M.L. Lake, *Diamond and Related Materials* 3(10), 1243-1248 (1994).

87. X. Wang and D.D.L. Chung, *Smart Mater. Struct.* 4, 363-367 (1995).

88. X. Shui and D.D.L. Chung, *Smart Mater. Struct.* 6, 102-105 (1997).

89. P.-W. Chen and D.D.L. Chung, *ACI Mater. J.* 93(4), 341-350 (1996).

90. P.-W. Chen and D.D.L. Chung, *Composites* 27B, 11-23 (1996).

91. P.-W. Chen and D.D.L. Chung, *Smart Mater. Struct.* 2, 22-30 (1993).

92. X. Shui and D.D.L. Chung, *Smart Mater. Struct.* 5, 243-246 (1996).

93. V. Chellappa, Z.W. Chiou and B.Z. Jang, *J. Mater. Sci.* 30(17), 4263-4272 (1995).

94. E.E. Hucke, U.S. Patent 3,859,421 (1975)

95. J. Wang, *Electrochim. Acta* 26, 1721-1726 (1981).

96. A.N. Strohl and D.J. Curran, *Anal. Chem.* 51, 1050-1053 (1979).

97. W.J. Blaedel and J. Wang, *Anal. Chem.* 52(11), 1697-1700 (1980).

98. I.C. Agarwal, A.M. Rochon, H.D. Gesser and A.B. Sparling, *Water Res.* 18, 227-232 (1984).

99. A.P. Sylwester, J.H. Aubert, P.B. Rand, C. Arnold, Jr. and L.R. Clough, *Polymer. Mater. Sci. Eng.* 57, 113-117 (1987).

100. A.P. Sylwester and R.L. Clough, *Synth Met.* 29(2-3), 253-258 (1989).

101. Y. Oren and A. Soffer, *Electrochim. Acta* 28, 1649-1654 (1983).

102. C. Lestrade, P.Y. Guyomar and M. Astruc, *Environ. Technol. Lett.* 2, 409 (1981).

103. X. Shui and D.D.L. Chung, *Carbon* 34(6), 811-814 (1996); 34(9), 1162 (1996).

104. M.M. Dubinin, N.S. Polyakov and G.A. Petukhova, *Adsorption Science Tech.* 10(1-4), 17-26 (1993).

105. H.-L. Chiang, P.C. Chiang and J.H. You, *Toxicol. Environ. Chem.* 47(1-2), 97-108 (1995).

106. Y. Takeuchi and T. Itoh, *Sep. Technol.* 3(3), 168-175 (1993).

107. J. Alcaniz-Monge, D. Cazorla-Amoros, A. Linares-Solano, S. Yoshida and A. Oya, *Carbon* 32(7), 1277-1283 (1994).

108. R. Ghosal, D.J. Kaul, U. Boes, D. Sanders, D.M. Smith and A. Maskara, Materials Research Society Symp. Proc., 371, 413-423 (1995).

109. A.W.P. Fung, Z.H. Wang, K. Lu, M.S. Dresslehaus and R.W. Pekala, *J. Mater. Res.* 8(8), 1875-1885 (1993).

110. R. Ströbel, L. Jörissen, T. Schliermann, V. Trapp, W. Schütz, K. Bohmhammel, G. Wolf and J. Garche, *J. Power Sources* 84(2), 221-224 (1999).

111. W. Lu and D.D.L. Chung, *Carbon* 35(3), 427-430 (1997).

112. M. Kim, N.M. Rodriguez and R.T.K. Baker, Mater. Res. Soc. Symp. Proc., 368, 99-104 (1995).

113. N.M. Rodriguez, M.-S. Kim and R.T.K. Baker, *J. Phys. Chem.* 98(50), 13108-13111 (1994).

114. C. Park and R.T.K. Baker, Mater. Res. Soc. Symp. Proc., 497, 145-150 (1998).

115. Chambers, T. Nemes, N.M. Rodriguez and R.T.K. Baker, *J. Phys. Chem.* 102, 2251-2258 (1998).

116. C. Park, N.M. Rodriguez and R.T.K. Baker, *Mater. Res. Soc. Symp. Proc.*, 454, 21-26 (1997).

117. C. Park, R.T.K. Baker and N.M. Rodriguez, Coke Formation and Mitigation Preprints — American Chemical Society, Div. Petroleum Chem., 40(4), 646-648 (1995).

118. J.W. Geus, A.J. Van Dillen and M.S. Hoogenraad, *Mater. Res. Soc. Symp. Proc.*, 368, 87-98 (1995).

11 Improving Cement-Based Materials by Using Silica Fume

CONTENTS

SYNOPSIS The effects of silica fume as an admixture in cement-based materials are reviewed in terms of the mechanical properties, vibration damping capacity, freeze-thaw durability, abrasion resistance, shrinkage, air void content, density, permeability, steel rebar corrosion resistance, alkali-silica reactivity reduction, chemical attack resistance, bond strength to steel rebar, creep rate, coefficient of thermal expansion, specific heat, thermal conductivity, fiber dispersion, and workability. The effects of silane treatment of the silica fume and of the use of silane as an additional admixture are also addressed.

11.1 INTRODUCTION

Cement-based materials such as concrete have long been used for civil infrastructure such as highways, bridges, and buildings. However, the deterioration of the civil infrastructure across the U.S. has led to the realization that cement-based materials must be improved in terms of their properties and durability. The fabrication of these materials involves mixing cement, water, aggregates, and other additives (called admixtures). The use of admixtures is a convenient way of improving cement-based materials. Techniques involving special mixing, casting, or curing procedures tend to be less attractive because of the need for special equipment in the field. An admixture that is particularly effective is silica fume,[1-4] although there are others such as latex and short fibers that are more expensive. This chapter focuses on the use of silica fume to improve cement-based materials.

Silica fume is very fine noncrystalline silica generated by electric arc furnaces as a by-product of the fabrication of metallic silicon or ferrosilicon alloys. It is a powder with particles having diameters 100 times smaller than Portland cement, i.e., mean particle size between 0.1 and 0.2 μm. The SiO_2 content ranges from 85 to 98%. Silica fume is pozzolanic, meaning it is reactive, like volcanic ash.

The property improvements needed for cement-based materials include increases in strength, modulus, and ductility; decrease in the drying shrinkage; decrease in the permeability to liquids and chloride ions; and increase in the durability to freeze-thaw temperature cycling.

Silica fume used as an admixture in a concrete mix has significant effects on the properties of the resulting material. These effects pertain to the strength, modulus, ductility, vibration damping capacity, sound absorption, abrasion resistance, air void content, shrinkage, bonding strength with reinforcing steel, permeability, chemical attack resistance, alkali-silica reactivity reduction, corrosion resistance of embedded steel reinforcement, freeze-thaw durability, creep rate, coefficient of thermal expansion (CTE), specific heat, thermal conductivity, and degree of fiber dispersion in mixes containing short microfibers. In addition, silica fume addition degrades the workability of the mix.

The data given in this chapter to illustrate the effects of silica fume are all for silica fume from Elkem Materials Inc. (EMS 965) used in the amount of 15% by weight of cement. The cement is Portland cement (Type I) from Lafarge Corp. Curing is in air at room temperature with a relative humidity of 100% for 28 days. The water-reducing agent, if used, is a sodium salt of a condensed naphthalenesulphonic acid from Rohm and Haas Co. (TAMOL SN®).

11.2 WORKABILITY

Silica fume causes workability and consistency losses,[5-18] which are barriers against proper utilization of silica fume concrete. However, the consistency of silica fume mortar is greatly enhanced by using either silane-treated silica fume or silane as an additional admixture.[19,20] The effectiveness of silane for cement is due to the reactivity of its molecular ends with –OH groups, which are present on the surface of both silica and cement.

TABLE 11.1
Workability of Mortar Mix (The water/cement ratio was 0.35.)

Silica Fume	Water-Reducing Agent/Cement	Slump (mm)
Plain	0%	*
With untreated silica fume	0%	150
With untreated silica fume	1%	186
With untreated silica fume	2%	220
With treated silica fume	0%	194
With treated silica fume	0.2%	215
With silane and untreated silica fume	0%	197
With silane and untreated silica fume	0.2%	218

* Too large to be measured.

Table 11.1[19,20] shows that silane introduction using either the coating or admixture method causes the silica fume mortar mix to increase in workability (slump). With silane (by either method) and no water-reducing agent, the workability of silica fume mortar mix is better than that of the mix with as-received silica fume and water-reducing agent in the amount of 1% by weight of cement. With silane (by either method) and water-reducing agent in the amount of 0.2% by weight of cement, the workability is almost as good as the mix with as-received silica fume and water-reducing agent in the amount of 2% by weight of cement.

The increase in workability due to silane introduction is because of the improved wettability of silica fume by water. The improved wettability is expected from the hydrophylic nature of the silane molecule. Silane treatment involves formation of a silane coating on the surface of the silica fume; it does not cause surface roughening.[21]

11.3 MECHANICAL PROPERTIES

The most well-known effect of silica fume is the increase in strength,[17,18,20,22-52] including compressive strength,[17,20,53-81] tensile strength,[20,60,63,71,72,82] and flexural strength.[59,66,72,83,84] The strengthening is due to the pozzolanic activity of silica fume causing improved strength of the cement paste;[36,79] the increased density of mortar or concrete resulting from the fineness of silica fume and the consequent efficient reaction to form hydration products, which fill the capillaries between cement and aggregate;[50] the refined pore structure;[55,76] and the microfiller effect of silica fume.[76,77] In addition, the modulus is increased.[20,44,48,53,73,85] These effects are also partly due to the densification of the interfacial zone between paste and aggregate.[85]

As shown in Table 11.2[20] for cement pastes at 28 days of curing, the tensile strength, tensile ductility, compressive strength, and compressive modulus are increased, and the compressive ductility is decreased by the addition of silica fume that has not been surface treated. The tensile strength and compressive strength are further increased, and the compressive ductility further decreased, when silane-

TABLE 11.2
Mechanical Properties of Cement Pastes at 28 Days of Curing

	Plain	With Untreated Silica Fume[†]	With Treated Silica Fume[†]	With Silane[*] and Untreated Silica Fume[†]
Tensile strength (MPa)	0.91 ± 0.02	1.53 ± 0.06	2.04 ± 0.06	2.07 ± 0.05
Tensile modulus (GPa)	11.2 ± 0.24	10.2 ± 0.7	11.5 ± 0.6	10.9 ± 0.5
Tensile ductility (%)	0.0041 ± 0.00008	0.020 ± 0.0004	0.020 ± 0.0004	0.021 ± 0.0004
Compressive strength (MPa)	57.9 ± 1.8	65.0 ± 2.6	77.3 ± 4.1	77.4 ± 3.7
Compressive modulus (GPa)	2.92 ± 0.07	13.6 ± 1.4	10.9 ± 1.8	15.8 ± 1.6
Compressive ductility (%)	1.72 ± 0.04	0.614 ± 0.023	0.503 ± 0.021	0.474 ± 0.015

[*] 0.2% by weight of cement
[†] 15% by weight of cement

treated silica fume is used. On the other hand, the tensile modulus is not affected by the silica fume addition. The use of both silane and untreated silica fume enhances the tensile strength, compressive strength, and compressive modulus, but decreases the compressive ductility relative to the paste with untreated silica fume and no silane. The effects of using treated silica fume and of using the combination of silane and untreated silica fume are quite similar, except that the compressive modulus is higher and the compressive ductility lower for the latter because of the network of covalent silane coupling among the silica fume particles.

The use of silane as an admixture added directly into the cement mix involves slightly more silane material but less processing cost than the use of silane as a coating on silica fume. Both methods result in increases in the tensile and compressive strengths. The network attained by the admixture method does not result from the silane coating method due to the localization of the silane in the coating, which nevertheless provides chemical coupling between silica fume and cement. The network, which is formed from the hydrolysis and polymerization reaction of silane during the hydration of cement, also causes the ductility to decrease.

11.4 VIBRATION DAMPING CAPACITY

Vibration reduction is valuable for hazard mitigation, structural stability, and structural performance improvement. Effective vibration reduction requires both damping capacity and stiffness. Silica fume is effective for enhancing both damping capacity and stiffness.[20,86-88]

As shown in Table 11.3,[20] the vibration damping capacity, as expressed by the loss tangent under dynamic 3-point flexural loading at 0.2 Hz, is significantly increased by the addition of silica fume that has not been surface treated. The use of silane-treated silica fume increases the loss tangent slightly beyond the value attained with untreated silica fume. The use of silane and untreated silica fume as two admixtures decreases the loss tangent to a value below that attained by using untreated silica fume alone, but still above that for plain cement paste.

TABLE 11.3
Dynamic Flexural Properties of Cement Pastes at a Flexural (3-Point Bending) Loading Frequency of 0.2 Hz

	Loss Tangent (tan δ, ±0.002)	Storage Modulus (GPa, ±0.03)	Loss Modulus (GPa, ±0.02)
Plain	0.035	1.91	0.067
With untreated silica fume[†]	0.082	12.71	1.04
With treated silica fume[†]	0.087	16.75	1.46
With silane[*] and untreated silica fume[†]	0.055	17.92	0.99

[*] 0.2% by weight of cement
[†] 15% by weight of cement

The ability of silica fume to enhance damping capacity is due to the large area of the interface between silica fume particles and the cement matrix, and the contribution of interface slippage to energy dissipation. Although the pozzolanic nature of silica fume makes the interface rather diffuse, the interface still contributes to damping. The silane covalent coupling introduced by the silane surface treatment of silica fume can move during vibration, thus providing another mechanism for damping and enhancing the loss tangent. The network introduced by the use of silane and untreated silica fume as two admixtures restricts movement, therefore reducing the damping capacity relative to the case with untreated silica fume alone. Nevertheless, the use of the two admixtures enhances the damping capacity relative to plain cement paste, as even less movement is possible in plain cement paste.

The storage modulus (Table 11.3) is much increased by the addition of untreated silica fume, is further increased by the use of silane-treated silica fume, and is still further increased by the use of silane and untreated silica fume as two admixtures. The increase in storage modulus upon addition of untreated silica fume is attributed to the high modulus of silica compared to the cement matrix. The enhancement of the storage modulus by the use of silane-treated silica fume is due to the chemical coupling provided by the silane between silica fume and cement. The further enhancement of the storage modulus by the use of silane and untreated silica fume as two admixtures is due to the network of covalent coupling among the silica fume particles.

The loss modulus (Table 11.3) is the product of the loss tangent and the storage modulus. As vibration reduction requires both damping and stiffness, both loss tangent and storage modulus should be high for effective vibration reduction. Hence, the loss modulus serves as an overall figure of merit for vibration reduction ability. The loss modulus is much increased by the addition of untreated silica fume, and is further increased by the use of silane-treated silica fume. However, the use of silane and untreated silica fume as two admixtures decreases the loss modulus to a value below the paste with untreated silica fume alone because of the decrease in the loss tangent. As a result, the use of silane-treated silica fume yields the highest value of the loss modulus.

11.5 SOUND ABSORPTION

Sound or noise absorption is useful for numerous structures, such as pavement overlays and noise barriers. The addition of silica fume to concrete improves sound absorption ability.[88] The effect is related to the increase in vibration damping capacity.

11.6 FREEZE-THAW DURABILITY

Freeze-thaw durability refers to the ability to withstand changes between temperatures above 0°C and those below 0°C. Due to the presence of water, which undergoes freezing and thawing, concrete tends to degrade upon such temperature cycling. Air voids (called air entrainment) are used as cushions to accommodate the changes in volume, thereby enhancing freeze-thaw durability.

The addition of silica fume to mortar improves freeze-thaw durability,[89-92] in spite of the poor air-void system.[92] However, the use of air entrainment is still recommended.[93-97] The addition of silica fume also reduces scaling.[98]

11.7 ABRASION RESISTANCE

The addition of untreated silica fume to mortar increases abrasion resistance,[48,98,99] as shown by the depth of wear decreasing from 1.07 to 0.145 mm (as tested using ASTM C944-90a, Rotating-Cutter Method).[99] Abrasion resistance is further improved by using acid-treated silica fume.[100]

11.8 SHRINKAGE

The hydration reaction that occurs during the curing of cement causes shrinkage, called "autogenous shrinkage," which is accompanied by a decrease in the relative humidity within the pores. If curing is conducted in an open atmosphere, as is often the case, additional shrinkage occurs due to the movement of water through the pores to the surface, and the loss of water on the surface by evaporation. The overall shrinkage that occurs is known as the "drying shrinkage," which is the shrinkage that is practically important.

Drying shrinkage can cause cracking and prestressing loss.[101] The addition of untreated silica fume to cement paste decreases drying shrinkage[20,34,101-105] (Table 11.4). This desirable effect is partly due to the reduction of the pore size and connectivity of the voids, and partly due to the prestressing effect of silica fume, which restrains shrinkage. The use of silane-treated silica fume in place of untreated silica fume further decreases drying shrinkage because of the hydrophylic character of the silane-treated silica fume, and the formation of chemical bonds between silica fume particles and cement.[20] The use of silane and untreated silica fume as two admixtures also decreases the drying shrinkage, but not as significantly as the use of silane-treated silica fume.[20] However, silica fume has also been reported to increase drying shrinkage,[34,106,107] and the restrained shrinkage crack width is increased by silica fume addition.[108]

TABLE 11.4
Drying Shrinkage Strain (10^{-4}, ±0.015)
of Cement Pastes at 28 Days

Plain	4.98
With untreated silica fume	4.41
With treated silica fume	4.18
With silane and untreated silica fume	4.32

Due to the pozzolanic nature of silica fume, silica fume addition increases the autogenous shrinkage, as well as the autogenous relative humidity change.[109,110] These effects are undesirable, as they may cause cracking if the deformation is restrained.

Carbonization refers to the chemical reaction between CO_2 and cement, as made possible by the in-diffusion of CO_2 gas. This reaction causes shrinkage, called "carbonization shrinkage." Due to the effect of silica fume addition on the pore structure, which affects the in-diffusion, carbonization shrinkage may be avoided by the addition of silica fume.[111]

Concrete exposed to hot climatic conditions soon after casting is particularly prone to plastic shrinkage cracking,[112] which is primarily due to the development of tensile capillary pressure during drying. Silica fume addition increases the plastic shrinkage[113] because of the high tensile capillary pressure resulting from the high surface area of the silica fume particles.

11.9 AIR VOID CONTENT AND DENSITY

The air void content of cement paste (Table 11.5) is increased by the addition of untreated silica fume.[20,114] Along with this effect is a decrease in density (Table 11.6). Both effects are related to the reduction in drying shrinkage. The introduction of silane by either coating or admixture decreases the air void content, but the value is still higher than that of plain cement paste. The use of the admixture method of silane introduction increases the density to a value almost as high as that of plain cement paste because of the network of covalent coupling among the silica fume particles.[20] On the other hand, the air void content of concrete is decreased by silica fume addition,[8,17] probably because of the densification of the paste-aggregate interface.[85]

11.10 PERMEABILITY

The permeability of chloride ions in concrete is decreased by the addition of untreated silica fume.[5,8,17,22,25,30-32,46,47,50,57,59,78,103,115-138] Related to this effect is the decrease in water absorptivity. Both effects are due to the microscopic pore structure resulting from the calcium silicate hydrate (CSH) formed upon the pozzolanic reaction of silica fume with free lime during the hydration of concrete.[139-163]

**TABLE 11.5
Air Void Content (%, ±0.02)
of Cement Pastes**

Plain	2.32
With untreated silica fume	3.73
With treated silica fume	3.26
With silane and untreated silica fume	3.19

**TABLE 11.6
Density (g/cm³, ±0.02) of Cement Pastes**

Plain	2.01
With untreated silica fume	1.72
With treated silica fume	1.73
With silane and untreated silica fume	1.97

11.11 STEEL REBAR CORROSION RESISTANCE

The addition of untreated silica fume to steel-reinforced concrete enhances the corrosion resistance of the reinforcing steel.[163-177] This is related to the decrease in permeability.

11.12 ALKALI-SILICA REACTIVITY REDUCTION

Alkali-silica reactivity refers to the reactivity of silica (present in most aggregates) and alkaline ions (present in cement). It is detrimental due to the expansion caused by the reaction product. This reactivity is reduced by the addition of silica fume[178-195] because of the effectiveness of silica fume in removing alkali from the pore solution,[184,185,190] reducing the alkali ion (Na^+, K^+, OH^-) concentrations in the pore solution,[193,195] and retarding the transportation of alkalis to reaction sites.[192] However, silica fume with coarse particles or undispersed agglomerates can induce distress related to alkali-silica reactivity.[196]

11.13 CHEMICAL ATTACK RESISTANCE

The addition of untreated silica fume to concrete enhances chemical attack resistance[197-213] whether the chemical is acid, sulfate, chloride, etc. This effect is related to the decrease in permeability.

11.14 BOND STRENGTH TO STEEL REBAR

The addition of untreated silica fume to concrete increases the shear bond strength between concrete and steel rebar.[214-220] This effect is mainly due to the reduced porosity and thickness of the transition zone adjacent to the steel, thereby improving

TABLE 11.7
Specific Heat (J/g.K, ±0.001) of Cement Pastes

Plain	0.736
With untreated silica fume	0.782
With treated silica fume	0.788
With silane and untreated silica fume	0.980

the adhesion-type bond at small slip levels.[217-220] The combined use of silica fume and methylcellulose (0.4% by weight of cement) gives even higher bond strength because of the surfactant role of methylcellulose.[218,221]

11.15 CREEP RATE

The addition of untreated silica fume to cement paste decreases the compressive creep rate at 200°C from 1.3×10^{-5} to 2.4×10^{-6} min^{-1}.[114] The creep resistance is consistent with the high storage modulus (Table 11.3), which remains much higher than that of plain cement paste, up to at least 150°C.[114] However, silica fume increases the early age tensile creep, which provides a mechanism to relieve some of the restraining stress that develops because of autogenous shrinkage.[222]

11.16 COEFFICIENT OF THERMAL EXPANSION

The CTE is reduced by the addition of untreated silica fume.[114] This is consistent with high modulus and creep resistance.

11.17 SPECIFIC HEAT

A high value of specific heat is valuable for improving the temperature stability of a structure, and to retain heat in a building. Specific heat (C_p, Table 11.7) is increased by the addition of untreated silica fume.[114] The use of silane-treated silica fume in place of untreated silica fume further increases specific heat, though only slightly.[223] The effect of untreated silica fume is due to the slippage at the interface between silica fume and cement. The effect of the silane treatment is because of the contribution of the movement of the covalent coupling between silica fume particles and cement. The use of silane and untreated silica fume as two admixtures greatly increases specific heat due to the network of covalent coupling among the silica fume particles contributing to phonons.[20]

11.18 THERMAL CONDUCTIVITY

Concrete of low thermal conductivity is useful for the thermal insulation of buildings. On the other hand, concrete of high thermal conductivity is useful for reducing temperature gradients in structures. The thermal stresses that result from temperature gradients may cause mechanical property degradation, and even warpage in the

TABLE 11.8
Thermal Conductivity (W/m.K, ±0.07)
of Cement Pastes

Plain	0.53
With untreated silica fume	0.35
With treated silica fume	0.33
With silane and untreated silica fume	0.61

structure. Bridges are among the structures that encounter temperature differentials between their top and bottom surfaces. In contrast to buildings, which also encounter temperature differentials, bridges do not need thermal insulation. Therefore, concrete of high thermal conductivity is desirable for bridges and related structures.

Thermal conductivity (Table 11.8) is decreased by the addition of untreated or silane treated silica fume[20,114,223] due to the interface between silica fume particles and cement acting as a barrier against heat conduction. Thermal conductivity is increased by the use of silane and untreated silica fume as two admixtures[20] because of the network of covalent coupling enhancing heat conduction through phonons.

11.19 FIBER DISPERSION

Short microfibers, such as carbon, glass, polypropylene, steel, and other fibers, are used as an admixture in concrete to enhance the tensile and flexural properties, and to decrease drying shrinkage. Effective use of the fibers, which are used in very small quantities, requires good dispersion. The addition of untreated silica fume to microfiber-reinforced cement increases the degree of fiber dispersion due to the fine silica fume particles helping the mixing of the microfibers.[224-269] In addition, silica fume improves the structure of the fiber-matrix interface, reduces the weakness of the interfacial zone, and decreases the number and size of cracks.[269]

11.20 CONCLUSION

The use of silica fume as an admixture in cement-based materials increases the tensile strength, compressive strength, compressive modulus, flexural modulus, and tensile ductility, but decreases compressive ductility. In addition, it enhances freeze-thaw durability, vibration damping capacity, abrasion resistance, bond strength with steel rebars, chemical attack resistance, and corrosion resistance of reinforcing steel. Furthermore, it decreases alkali-silica reactivity, drying shrinkage, permeability, creep rate, and coefficient of thermal expansion. It also increases specific heat and decreases thermal conductivity, though thermal conductivity is increased if silica fume is used with silane. Silica fume addition also increases air void content, decreases density, enhances dispersion of microfibers, and decreases workability.

The use of silane-treated silica fume in place of untreated silica fume increases consistency, tensile strength, and compressive strength, but decreases compressive ductility. Furthermore, the silane treatment increases damping capacity and specific heat, and decreases drying shrinkage and air void content.

The use of silane and untreated silica fume as two admixtures, relative to the use of silane-treated silica fume, increases compressive modulus, but decreases compressive ductility and damping capacity. It also decreases air void content and increases density, specific heat, and thermal conductivity.

REFERENCES

1. M.D. Luther and P.A. Smith, *Proc. Eng. Foundation Conf.* 75-106 (1991).
2. V.M. Malhotra, *Concr. Int.* 15(4), 23-28 (1993).
3. M.D. Luther, *Concr. Int.* 15(4), 29-33 (1993).
4. J. Wolsiefer and D.R. Morgan, *Concr. Int.* 15(4), 34-39 (1993).
5. M.D.A. Thomas, M.H. Shehata, S.G. Shashiprakash, D.S. Hopkins and K. Cail, *Cem. Concr. Res.* 29(8), 1207-1214 (1999).
6. J. Punkki, J. Golaszewski and O.E. Gjorv, *ACI Mater. J.* 93(5), 427-431 (1996).
7. R.P. Khatri, V. Sirivivatnanon and W. Gross, *Cem. Concr. Res.* 25(1), 209-220 (1995).
8. Z. Bayasi and R. Abitaher, *Concr. Int.* 14(4), 35-37 (1992).
9. O.E. Gjorv, *Concr. Int.* 20(9), 57-60 (1998).
10. H. El-Didamony, A. Amer, M. Heikal and M. Shoaib, *Ceramics-Silikaty* 43(1), 29-33 (1999).
11. H. El-Didamony, A. Amer and M. Heikal, *Ceramics-Silikaty* 42(4), 171-176 (1998).
12. M. Nehdi, S. Mindess and P.-C. Aitcin, *Cem. Concr. Res.* 28(5), 687-697 (1998).
13. F.E. Amparano and Y. Xi, *ACI Mater. J.* 95(6), 695-703 (1998).
14. M. Nehdi and S. Mindess, *Transp. Res. Rec.* (1574), 41-48 (1996).
15. L. Kucharska and M. Moczko, *Adv. Cem. Res.* 6(24), 139-145 (1994).
16. P. Rougeron and P.-C. Aitcin, *Cem. Concr. Aggregates* 16(2), 115-124 (1994).
17. Z. Bayasi and J. Zhou, *ACI Mater. J.* 90(4), 349-356 (1993).
18. F. Collins and J.G. Sanjayan, *Cem. Concr. Res.* 29(3), 459-462 (1999).
19. Y. Xu and D.D.L. Chung, *Cem. Concr. Res.* 29(3), 451-453 (1999).
20. Y. Xu and D.D.L. Chung, *Cem. Concr. Res.* 30(8), 1305-1311 (2000).
21. Y. Xu and D.D.L. Chung, *Composite Interfaces* 7(4), 243-256 (2000).
22. S.Y.N. Chan and X. Ji, *Cem. Concr. Compos.* 21(4), 293-300 (1999).
23. V.G. Papadakis, *Cem. Concr. Res.* 29(1), 79-86 (1999).
24. M.D.A. Thomas, K. Cail and R.D. Hooton, *Can. J. Civ. Eng.* 25(3), 391-400 (1998).
25. H.A. Toutanji, *Adv. Cem. Res.* 10(3), 135-139 (1998).
26. R. Lewis, *Concrete* 32(5), 19-20,22 (1998).
27. R. Breitenbuecher, *Mater. Struct.* 31(207), 209-215 (1998).
28. B. Persson, *Adv. Cem. Based Mater.* 7(3-4), 139-155 (1998).
29. W.H. Dilger and S.V.K.M. Rao, *Pci J.* 42(4), 82-96 (1997).
30. K.H. Khayat, M. Vachon and M.C. Lanctot, *ACI Mater. J.* 94(3), 183-192 (1997).
31. S.A. El-Desoky and I.A. Ibrahim, *Eur. J. Control* 43(5), 919-932 (1996).
32. M.L. Allan and L.E. Kukacka, *ACI Mater. J.* 93(6), 559-568 (1996).
33. Y. Li, B.W. Langan and M.A. Ward, *Cem., Concr. Aggregates* 18(2), 112-117 (1996).
34. M.N. Haque, *Cem. Concr. Compos.* 18(5), 333-342 (1996).
35. K. Wiegrink, S. Marikunte and S.P. Shah, *ACI Mater. J.* 93(5), 409-415 (1996).
36. M. Kessal, P.-C. Nkinamubanzi, A. Tagnit-Hamou and P.-C. Aitcin, *Cem. Concr. Aggregates* 18(1), 49-54 (1996).

37. V. Lilkov and V. Stoitchkov, *Cem. Concr. Res.* 26(7), 1073-1081 (1996).

38. M. Kessal, M. Edwards-Lajnef, A. Tagnit-Hamou and P.-C. Aitcin, *Can. J. Civil Eng.* 23(3), 614-620 (1996).

39. W.H. Dilger, A. Ghali, S.V.K.M. Rao, *Pci J.* 41(2), 68-89 (1996).

40. S.A.A. El-Enein, M.F. Kotkata, G.B. Hanna, M. Saad, and M.M.A. El Razek, *Cem. Concr. Res.* 25(8), 1615-1620 (1995).

41. S. Wild, B.B. Sabir and J.M. Khatib, *Cem. Concr. Res.* 25(7), 1567-1580 (1995).

42. K.G. Babu and P.V.S. Prakash, *Cem. Concr. Res.* 25(6), 1273-1283 (1995).

43. M. Collepardi, S. Monosi and P. Piccioli, *Cem. Concr. Res.* 25(5), 961-968 (1995).

44. J.C. Walraven, *Betonwerk und Fertigteil-Technik* 60(11), 7 pp (1994).

45. F. Papworth and R. Ratcliffe, *Concr. Int.* 16(10), 39-44 (1994).

46. K. Torii and M. Kawamura, *Cem. Concr. Compos.* 16(4), 279-286 (1994).

47. C. Ozyildirim and W.J. Halstead, *ACI Mater. J.* 91(6), 587-594 (1994).

48. S.A. Khedr and M.N. Abou-Zeid, *J. Mater. Civil Eng.* 6(3), 357-375 (1994).

49. P. Fidiestol, *Concr. Int.* 15(11), 33-36 (1993).

50. L. Rocole, *Aberdeen's Concr. Constr.* 38(6), 441-442 (1993).

51. G. Ozyildirim, *Concr. Int.* 15(1), 33-38 (1993).

52. M. Moukwa, B.G. Lewis, S.P. Shah and C. Ouyang, *Cem. Concr. Res.* 23(3), 711-723 (1993).

53. B. Ma, J. Li and J. Peng, *J. Wuhan Univ. Technol.* 14(2), 1-7 (1999).

54. K. Tan and X. Pu, *Cem. Concr. Res.* 28(12), 1819-1825 (1998).

55. L. Bagel, *Cem. Concr. Res.* 28(7), 1011-1020 (1998).

56. C.E.S. Tango, *Cem. Concr. Res.* 28(7), 969-983 (1998).

57. M. Lachemi, G. Li, A. Tagnit-Hamou and P.-C. Aitcin, *Concr. Int.* 20(1), 59-65 (1998).

58. S.H. Alsayed, *ACI Mater. J.* 94(6), 472-477 (1997).

59. Salas, R. Gutierrez and S. Delvasto, *J. Resour. Manage. Technol.* 24(2), 74-78 (1997).

60. J. Li and T. Pei, *Cem. Concr. Res.* 27(6), 833-837 (1997).

61. B.B. Sabir, *Mag. Concr. Res.* 49(179), 139-146 (1997).

62. V. Waller, P. Naproux and F. Larrard, *Bull. Liaison Lab. Ponts Chaussees* (208), 53-65 (1997).

63. H.A. Toutanji and T. El-Korchi, *Cem. Concr. Aggregates* 18(2), 78-84 (1996).

64. S. Iravani, *ACI Mater. J.* 93(5), 416-426 (1996).

65. M.H. Zhang, R. Lastra and V.M. Malhotra, *Cem. Concr. Res.* 26(6), 963-977 (1996).

66. R. Gagne, A. Boisvert and M. Pigeon, *ACI Mater. J.* 93(2), 111-120 (1996).

67. S.L. Mak and K. Torii, *Cem. Concr. Res.* 25(8), 1791-1802 (1995).

68. B.B. Sabir, *Mag. Concr. Res.* 47(172), 219-226 (1995).

69. H.A. Toutanji and T. El-Korchi, *Cem. Concr. Res.* 25(7), 1591-1602 (1995).

70. S. Ghosh and K.W. Nasser, *Can. J. Civil Eng.* 22(3), 621-636 (1995).

71. H. Marzouk and A. Hussein, *J. Mater. Civil Eng.* 7(3), 161-167 (1995).

72. E.H. Fahmy, Y.B.I. Shaheen and W.M. El-Dessouki, *J. Ferrocement* 25(2), 115-121 (1995).

73. F.P. Zhou, B.I.G. Barr and F.D. Lydon, *Cem. Concr. Res.* 25(3), 543-552 (1995).

74. J. Xie, A.E. Elwi and J.G. MacGregor, *ACI Mater. J.* 92(2), 135-145 (1995).

75. S.U. Al-Dulaijan, A.H.J. Al-Tayyib, M.M. Al-Zahrani, G. Parry-Jones and A.I. Al-Mana, *J. Am. Ceram. Soc.* 78(2), 342-346 (1995).

76. A. Goldman and A. Bentur, *Adv. Cem. Based Mater.* 1(5), 209-215 (1994).

77. A. Goldman and A. Bentur, *Cem. Concr. Res.* 23(4), 962-972 (1993).

78. K.J. Folliard, M. Ohta, E. Rathje and P. Collins, *Cem. Concr. Res.* 24(3), 424-432 (1994).

79. C. Xiaofeng, G. Shanglong, D. Darwin and S.L. McCabe, *ACI Mater. J.* 89(4), 375-387 (1992).

80. R.S. Ravindrarajah, Proc. Nondestr. Test. Concr. Elem. Struct., ASCE, pp. 115-126 (1992).

81. B.B. Sabir and K. Kouyiali, *Cem. Concr. Res.* 13(3), 203-208 (1991).

82. H.A. Toutanji, L. Liu and T. El-Korchi, *Mater. Struct.* 32(217), 203-209 (1999).

83. S. Sarkar, O. Adwan and J.G.L. Munday, *Struct. Eng.* 75(7), 115-121 (1997).

84. C. Tasdemir, M.A. Tasdemir, N. Mills, B.I.G. Barr and F.D. Lydon, *ACI Mater. J.* 96(1), 74-83 (1999).

85. M.G. Alexander and T.I. Milne, *ACI Mater. J.* 92(3), 227-235 (1995).

86. X. Fu, X. Li and D.D.L. Chung, *J. Mater. Sci.* 33, 3601-3605 (1998).

87. Y. Wang and D.D.L. Chung, *Cem. Concr. Res.* 28(10), 1353-1356 (1998).

88. M. Tamai and M. Tanaka, *Trans. Jpn. Concr. Inst.* 16, 81-88 (1994).

89. R. Quaresima, G. Scoccia, R. Volpe, F. Medici and C. Merli, Proc. 1993 Symp. Stabilization and Solidification of Hazardous, Radioactive, and Mixed Wastes: 3rd Volume, Vol. 1240, ASTM Special Technical Publication, pp. 135-146 (1996).

90. M. Sonebi and K.H. Khayat, *Can. J. Civil Eng.* 20(4), 650-659 (1993).

91. R.D. Hooton, *ACI Mater. J.,* 90(2), 143-151 (1993).

92. M. Lachemi, G. Li, A. Tagnit-Hamou and P.-C. Aitcin, *Concr. Int.* 20(1), 59-65 (1998).

93. V. Yogendran, B.W. Langan, M.N. Haque and M.A. Ward, *ACI Mater. J.* 84(2), 124-129 (1987).

94. P.-W. Chen and D.D.L. Chung, *Composites* 24(1), 33-52 (1993).

95. M.G. Kashi and R.E. Weyers, Proc. Sessions Related to Structural Materials at Structures Congress, Vol. 89, ASCE, pp. 138-148 (1989).

96. V.M. Malhotra, Proc. 2nd Int. Conf., Publication SP — American Concrete Institute 91, Vol. 2, pp. 1069-1094 (1986).

97. A. Durekovic, V. Calogovic and K. Popovic, *Cem. Concr. Res.* 19(2), 267-277 (1989).

98. B.B. Sabir, *Cem. Concr. Compos.* 19(4), 285-294 (1997).

99. Z.-Q. Shi and D.D.L. Chung, *Cem. Concr. Res.* 27(8), 1149-1153 (1997).

100. X. Li and D.D.L. Chung, *Cem. Concr. Res.* 28(4), 493-498 (1998).

101. V. Baroghel-Bouny and J. Godin, *Bull. Liaison Lab. Ponts Chaussees* (218), 39-48 (1998).

102. S.H. Alsayed, *Cem. Concr. Res.* 28(10), 1405-1415 (1998).

103. A. Lamontagne, M. Pigeon, R. Pleau and D. Beaupre, *ACI Mater. J.* 93(1), 69-74 (1996).

104. M.G. Alexander, *Adv. Cem. Res.* 6(22), 73-81 (1994).

105. F.H. Al-Sugair, *Mag. Concr. Res.* 47(170), 77-81 (1995).

106. B. Bissonnette and M. Pigeon, *Cem. Concr. Res.* 25(5), 1075-1085 (1995).

107. G.A. Rao, *Cem. Concr. Res.* 28(10), 1505-1509 (1998).

108. Z. Li, M. Qi, Z. Li, and B. Ma, *J. Mater. Civil Eng.* 11(3), 214-223 (1999).

109. O.M. Jensen and P.F. Hansen, *ACI Mater. J.* 93(6), 539-543 (1996).

110. O.M. Jensen and P.F. Hansen, *Adv. Cem. Res.* 7(25), 33-38 (1995).

111. B. Persson, *Cem. Concr. Res.* 28(7), 1023-1036 (1998).

112. T.A. Samman, W.H. Mirza and F.F. Wafa, *ACI Mater. J.* 93(1), 36-40 (1996).

113. R. Bloom and A. Bentur, *ACI Mater. J.* 92(2), 211-217 (1995).

114. X. Fu and D.D.L. Chung, *ACI Mater. J.* 96(4), 455-461 (1999).

115. Z. Liu and J.J. Beaudoin, *Cem. Concr. Res.* 29(7), 1085-1090 (1999).

116. T.-J. Zhao, J.-Q. Zhu and P.-Y. Chi, *ACI Mater. J.* 96(1), 84-89 (1999).

117. T.H. Wee, A.K. Suryavanshi and S.S. Tin, *Cem. Concr. Compos.* 21(1), 59-72 (1999).

118. R.K. Dhir and M.R. Jones, *Fuel* 78(2), 137-142 (1999).

119. P. Sandberg, L. Tang and A. Anderson, *Cem. Concr. Res.* 28(10), 1489-1503 (1998).

120. N. Gowripalan and H.M. Mohamed, *Cem. Concr. Res.* 28(8), 1119-1131 (1998).

121. C. Shi, J.A. Stegemann and R.J. Caldwell, *ACI Mater. J.* 95(4), 389-394 (1998).

122. S.L. Amey, D.A. Johnson, M.A. Miltenberger and H. Farzam, *ACI Struct. J.* 95(2), 205-214 (1998).

123. R.J. Detwiler, T. Kojundic and P. Fidjestol, *Concr. Int.* 19(8), 43-45 (1997).

124. S.L. Amey, D.A. Johnson, M.A. Miltenberger and H. Farzam, *ACI Struct. J.* 95(1), 27-36 (1998).

125. R.P. Khatri, V. Sirivivatnanon and L.K. Yu, *Mag. Concr. Res.* 49(180), 167-172 (1997).

126. R.P. Khatri, V. Sirivivatnanon and J.L. Yang, *Cem. Concr. Res.* 27(8), 1179-1189 (1997).

127. K. Tan and O.E. Gjorv, *Cem. Concr. Res.* 26(3), 355-361 (1996).

128. G.J.Z. Xu, D.F. Watt and P.P. Hudec, *Cem. Concr. Res.* 25(6), 1225-1236 (1995).

129. A.A. Ramezanianpour and V.M. Malhotra, *Cem. Concr. Compos.* 17(2), 125-133 (1995).

130. O.S.B. Al-Amoudi, M. Maslehuddin and Y.A.B. Abdul-Al, *Constr. Build. Mater.* 9(1), 25-33 (1995).

131. A. Delagrave, M. Pigeon and E. Revertegat, *Cem. Concr. Res.* 24(8), 1433-1443 (1994).

132. C. Ozyildirim, *Cem. Concr. Aggregates* 16(1), 53-56 (1994).

133. C. Ozyildirim, *ACI Mater. J.* 91(2), 197-202 (1994).

134. R.J. Detwiler, C.A. Fapohunda and J. Natale, *ACI Mater. J.* 91(1), 63-66 (1994).

135. C. Lobo and M.D. Cohen, *ACI Mater. J.* 89(5), 481-491 (1992).

136. N. Banthia, M. Pigeon, J. Marchand and J. Boisvert, *J. Mater. Civil Eng.* 4(1), 27-40 (1992).

137. B. Ma, Z. Li and J. Peng, *J. Wuhan Univ. Technol.* 13(4), 16-24 (1998).

138. P.S. Mangat and B.T. Molloy, *Mag. Concr. Res.* 47(171), 129-141 (1995).

139. C.S. Poon, L. Lam and Y.L. Wong, *J. Mater. Civil. Eng.* 11(3), 197-205 (1999).

140. H. Yan, W. Sun and H. Chen, *Cem. Concr. Res.* 29(3), 423-426 (1999).

141. D.R.G. Mitchell, I. Hinczak and R.A. Day, *Cem. Concr. Res.* 28(11), 1571-1584 (1998).

142. K.O. Kjellsen, O.H. Wallevik and L. Fjallberg, *Adv. Cem. Res.* 10(1), 33-40 (1998).

143. K. Vivekanandam and I. Patnaikuni, *Cem. Concr. Res.* 27(6), 817-823 (1997).

144. M. Saad, S.A. Abo-El-Enein, G.B. Hanna and M.F. Kotkata, *Cem. Concr. Res.* 26(10), 1479-1484 (1996).

145. K.A. Khalil, *Mater. Lett.* 26(4-5), 259-264 (1996).

146. J. Marchand, H. Hornain, S. Diamond, M. Pigeon and H. Guiraud, *Cem. Concr. Res.* 26(3), 427-438 (1996).

147. C. Tasdemir, M.A. Tasdemir, F.D. Lydon, and B.I.G. Barr, *Cem. Concr. Res.* 26(1), 63-68 (1996).

148. M.J. Aquino, Z. Li and S.P. Shah, *Adv. Cem. Based Mater.* 2(6), 211-223 (1995).

149. M. Cheyrezy, V. Maret and L. Frouin, *Cem. Concr. Res.* 25(7), 1491-1500 (1995).

150. D.P. Bentz and P.E. Stutzman, *Cem. Concr. Res.* 24(6), 1044-1050 (1994).

151. Y. Cao and R.J. Detwiler, *Cem. Concr. Res.* 25(3), 627-638 (1995).

152. P.-C. Aitcin, *Constr. Build. Mater.* 9(1), 13-17 (1995).

153. A. Durekovic, *Cem. Concr. Res.* 25(2), 365-375 (1995).

154. S. Mindess, L. Qu and M.G. Alexander, *Adv. Cem. Res.* 6(23), 103-107 (1994).

155. Q. Yu, M. Hu, J. Qian, X. Wang and C. Tao, *J. Wuhan Univ. Technol.* 9(2), 22-28 (1994).

156. K. Mitsui, Z. Li, D.A. Lange and S.P. Shah, *ACI Mater. J.* 91(1), 30-39 (1994).

157. M.D. Cohen, A. Goldman and W.-F. Chen, *Cem. Concr. Res.* 24(1), 95-98 (1994).
158. D.N. Winslow, M.D. Cohen, D.P. Bentz, K.A. Snyder and E.J. Garboczi, *Cem. Concr. Res.* 24(1), 25-37 (1994).
159. Z. Xu, P. Gu, P. Xie and J.J. Beaudoin, *Cem. Concr. Res.* 23(5), 1007-1015 (1993).
160. S.L. Sarkar, S. Chandra and L. Berntsson, *Cem. Concr. Compos.* 14(4), 239-248 (1992).
161. D.P. Bentz, P.E. Stutzman and E.J. Garboczi, *Cem. Concr. Res.* 22(5), 891-902 (1992).
162. X. Ping and J.J. Beaudoin, *Cem. Concr. Res.* 22(4), 597-604 (1992).
163. S.P. Shah, Z. Li and D.A. Lange, Proc. 9th Conf. Eng. Mechanics, pp. 852-855 (1992).
164. J. Zemajtis, R.E. Weyers and M.M. Sprinkel, *Transp. Res. Rec.* 57-59 (1999).
165. O.E. Gjorv, *ACI Mater. J.* 92(6), 591-598 (1995).
166. J.G. Cabrera, P.A. Claisse and D.N. Hunt, *Constr. Build. Mater.* 9(2), 105-113 (1995).
167. N.R. Jarrah, O.S.B. Al-Amoudi, M. Maslehuddin, O.A. Ashiru and A.I. Al-Mana, *Constr. Build. Mater.* 9(2), 97-103 (1995).
168. T. Lorentz and C. French, *ACI Mater. J.* 92(2), 181-190 (1995).
169. S.A. Khedr and A.F. Idriss, *J. Mater. Civil Eng.* 7(2), 102-107 (1995).
170. P.S. Mangat, J.M. Khatib and B.T. Molloy, *Cem. Concr. Compos.* 16(2), 73-81 (1994).
171. O.S.B. Al-Amoudi, A.R. Rasheeduzzafar, M. Maslehuddin and S.N. Abduljauwad, *Cem. Concr. Aggregates* 16(1), 3-11 (1994).
172. D. Whiting, *Transp. Res. Rec.* (1392), 142-148 (1993).
173. J.T. Wolsiefer, Sr., Proc. ASCE Natl. Conv. Expo., pp. 15-29 (1993).
174. A.R. Rasheeduzzafar, S.S. Al-Saadoun and A.S. Al-Gahtani, *ACI Mater. J.* 89(4), 337-366 (1992).
175. A.R. Rasheeduzzafar, *ACI Mater. J.* 89(6), 574-586 (1992).
176. C.K. Nmai, S.A. Farrington and G.S. Bobrowski, *Concr. Int.* 14(4), 45-51 (1992).
177. J. Hou and D.D.L. Chung, *Corros. Sci.* 42(9), 1489-1507 (2000).
178. V.S. Ramachandran, *Cem. Concr. Compos.* 20(2/3), 149-161 (1998).
179. S. Diamond, *Cem. Concr. Compos.* 19(5-6), 391-401 (1997).
180. H. Wang and J.E. Gillott, *Mag. Concr. Res.* 47(170), 69-75 (1995).
181. W. Wieker, R. Herr and C. Huebert, *Betonwerk und Fertigteil-Technik* 60(11), 86-91 (1994).
182. M. Berra, T. Mangialardi and A.E. Paolini, *Cem. Concr. Compos.* 16(3), 207-218 (1994).
183. M. Geiker and N. Thaulow, Proc. Eng. Foundation Conf. 123-136 (1991).
184. A. Shayan, G.W. Quick and C.J. Lancucki, *Adv. Cem. Res.* 5(20), 151-162 (1993).
185. J. Duchesne and M.A. Berube, *Cem. Concr. Res.* 24(2), 221-230 (1994).
186. P.P. Hudec and N.K. Banahene, *Cem. Concr. Compos.* 15(1-2), 21-26 (1993).
187. J.E. Gillot and H. Wang, *Cem. Concr. Res.* 23(4), 973-980 (1993).
188. N.R. Swamy and R.M. Wan, *Cem. Concr. Aggregates* 15(1), 32-49 (1993).
189. D.S. Lane, Mater. Eng. Congr., ASCE, pp. 231-244 (1992).
190. A.R. Rasheeduzzafar and S.E. Hussain, *Cem. Concr. Compos.* 13(3), 219-225 (1991).
191. C. Perry and J.E. Gillott, *Durability Build. Mater.* 3(2), 1985 (1985).
192. A.R. Rasheeduzzafar and H.S. Ehtesham, *Cem. Concr. Compos.* 13(3), 219-225 (1991).
193. B. Durand, J. Berard and R. Roux, *Cem. Concr. Res.* 20(3), 419-428 (1990).
194. G.J.Z. Xu, D.F. Watt and P.P. Hudec, *Cem. Concr. Res.* 25(6), 1225-1236 (1995).
195. M.-A. Berube, J. Duchesne and D. Chouinard, *Cem. Concr. Aggregates* 17(1), 26-33 (1995).
196. S. Diamond, *Cem. Concr. Compos.* 19(5-6), 391-401 (1997).
197. A.H. Ali, *Corros. Prev. Control* 46(3), 76-81 (1999).

198. M.S. Morsy, *Cem. Concr. Res.* 29(4), 603-606 (1999).

199. R.J. van Eijk and H.J.H. Brouwers, *Heron* 42(4), 215-229 (1997).

200. A.K. Tamimi, *Mater. Struct.* 30(197), 188-191 (1997).

201. F. Turker, F. Akoz, S. Koral and N. Yuzer, *Cem. Concr. Res.* 27(2), 205-214 (1997).

202. F.M. Kilinckale, *Cem. Concr. Res.* 27(12), 1911-1918 (1997).

203. N.D. Belie, V.D. Coster and D.V. Nieuwenburg, *Mag. Concr. Res.* 49(181), 337-344 (1997).

204. P. Jones, *Concrete* (London) 31(4), 12-13 (1997).

205. J.A. Daczko, D.A. Johnson and S.L. Amey, *Mater. Performance* 36(1), 51-56 (1997).

206. A. Delagrave, M. Pigeon, J. Marchand and E. Revertagat, *Cem. Concr. Res.* 26(5), 749-760 (1996).

207. A.A. Bubshait, B.M. Tahir and M.O. Jannadi, *Build. Res. Inf.* 24(1), 41-49 (1996).

208. F. Akoz, F. Turker, S. Koral and N. Yuzer, *Cem. Concr. Res.* 25(6), 1360-1368 (1995).

209. K. Torii and M. Kawamura, *Cem. Concr. Res.* 24(2), 361-370 (1994).

210. H.S. Shin and K.-S. Jun, C *J. Environ. Sci. Health* 30(3), :651-668 (1995).

211. V. Matte and M. Moranville, *Cem. Concr. Compos.* 21(1), 1-9 (1999).

212. M.J. Rudin, *Waste Management* 16(4), 305-311 (1996).

213. P.-W. Chen and D.D.L. Chung, *Composites* 27B, 269-274 (1996).

214. A. Yahia, K.H. Khayat and B. Benmokrane, *Mater. Struct.* 31(208), 267-274 (1998).

215. B.S. Hamad and S.M. Sabbah, *Mater. Struct.* 31(214), 707-713 (1998).

216. Z. Li, M. Xu and N.C. Chui, *Mag. Concr. Res.* 50(1), 49-57 (1998).

217. A. Mor, *ACI Mater. J.* 89(1), 76-82 (1992).

218. X. Fu and D.D.L. Chung, *ACI Mater. J.* 95(6), 725-734 (1998).

219. P.J.M. Monteiro, O.E. Gjorv and P.K. Mehta, *Cem. Concr. Res.* 19(1), 114-123 (1989).

220. O.E. Gjorv, P.J.M. Monteiro and P.K. Mehta, *ACI Mater. J.* 87(6), 573-580 (1990).

221. X. Fu and D.D.L. Chung, *ACI Mater. J.* 95(5), 601-608 (1998).

222. K. Kovler, S. Igarashi and A. Bentur, *Mater. Struct.* 32(219), 383-387 (1999).

223. Y. Xu and D.D.L. Chung, *Cem. Concr. Res.* 29(7), 1117-1121 (1999).

224. S.B. Park, E.S. Yoon and B.I. Lee, *Cem. Concr. Res.* 29(2), 193-200 (1999).

225. M. Pigeon and R. Cantin, *Cem. Concr. Compos.* 20(5), 365-375 (1998).

226. H. Toutanji, S. McNeil and Z. Bayasi, *Cem. Concr. Res.* 28(7), 961-968 (1998).

227. T.-J. Kim and C.-K. Park, *Cem. Concr. Res.* 28(7), 955-960 (1998).

228. H.A.D. Kirsten, *J. S. Afr. Inst. Min. Metall.* 98(2), 93-104 (1998).

229. C. Aldea, S. Marikunte and S.P. Shah, *Adv. Cem. Based Mater.* 8(2), 47-55 (1998).

230. A. Dubey and N. Banthia, *ACI Mater. J.* 95(3), 284-292 (1998).

231. L. Biolzi, G.L. Guerrini and G. Rosati, *Constr. Build. Mater.* 11(1), 57-63 (1997).

232. O. Eren and T. Celik, *Constr. Build. Mater.* 11(7-8), 373-382 (1997).

233. S.A. Austin, C.H. Peaston and P.J. Robins, *Constr. Build. Mater.* 11(5-6), 291-298 (1997).

234. E. Dallaire, P.-C. Aitcin and M. Lachemi, *Civ. Eng.* 68(1), 48-51 (1998).

235. R.J. Brousseau and G.B. Pye, *ACI Mater. J.* 94(4), 306-310 (1997).

236. S. Marikunte, C. Aldea and S.P. Shah, *Adv. Cem. Based Mater.* 5(3-4), 100-108 (1997).

237. Y.-W. Chan and V.C. Li, *Adv. Cem. Based Mater.* 5(1), 8-17 (1997).

238. R. Cantin and M. Pigeon, *Cem. Concr. Res.* 26(11), 1639-1648 (1996).

239. M.R. Taylor, F.D. Lydon and B.I.G. Barr, *Constr. Build. Mater.* 10(6), 445-450 (1996).

240. H.-C. Wu, V.C. Li, Y.M. Lim, K.F. Hayes and C.C. Chen, *J. Mater. Sci. Lett.* 15(19), 1736-1739 (1996).

241. A. Kumar and A.P. Gupta, *Exp. Mech.* 36(3), 258-261 (1996).

242. S. Wei, G. Jianming and Y. Yun, *ACI Mater. J.* 93(3), 206-212 (1996).

243. N. Banthia and C. Yan, *Cem. Concr. Res.* 26(5), 657-662 (1996).

244. A. Katz and A. Bentur, *Adv. Cem. Based Mater.* 3(1), 1-13 (1996).
245. P. Balaguru and A. Foden, *ACI Struct. J.* 93(1), 62-78 (1996).
246. P. Soroushian, F. Mirza and A. Alhozaimy, *ACI Mater. J.* 92(3), 291-295 (1995).
247. A. Katz and A. Bentur, *Cem. Concr. Compos.* 17(2), 87-97 (1995).
248. N. Banthia, N. Yan, C. Chan, C. Yan and A. Bentur, Materials Research Society Symposium Proceedings, Vol. 370, pp. 539-548 (1995).
249. A. Katz and V.C. Li, Materials Research Society Symposium Proceedings, Vol. 370, pp. 529-537 (1995).
250. A. Katz, V.C. Li and A. Kazmer, *J. Mater. Civil Eng.* 7(2), 125-128 (1995).
251. W.M. Boone, D.B. Clark and E.L. Theisz, Proc. Ports '95 Conference on Port Engineering and Development for the 21st Century, Vol. 2, pp. 1138-1147 (1995).
252. S.-I. Igarashi and M. Kawamura, *Proc. Jpn. Soc. Civ. Eng.* (502), Pt. 5-25, 83-92 (1994).
253. S. Marikunte and P. Soroushian, *ACI Mater. J.* 91(6), 607-616 (1994).
254. X. Lin, M.R. Silsbee, D.M. Roy, K. Kessler, and P.R. Blankenhorn, *Cem. Concr. Res.* 24(8), 1558-1566 (1994).
255. Anonymous, *Eng. News-Rec.* 233(2), 13 (1994).
256. N.M.P. Low and J.J. Beaudoin, *Cem. Concr. Res.* 24(5), 874-884 (1994).
257. N.M.P. Low and J.J. Beaudoin, *Cem. Concr. Res.* 24(2), 250-258 (1994).
258. A. Katz and A. Bentur, *Cem. Concr. Res.* 24(2), 214-220 (1994).
259. P. Balaguru and M.G. Dipsia, *ACI Mater. J.* 90(5), 399-405 (1993).
260. M.N.P. Low and J.J. Beaudoin, *Cem. Concr. Res.* 23(6), 1467-1479 (1993).
261. P.S. Mangat and G.S. Manarakis, *Mater. Struct.* 26(161), 433-440 (1993).
262. N.M.P. Low and J.J. Beaudoin, *Cem. Concr. Res.* 23(5), 1016-1028 (1993).
263. N.M.P. Low and J.J. Beaudoin, *Cem. Concr. Res.* 23(4), 905-916 (1993).
264. S.B. Park and B.I. Lee, *High Temp. High Pressures* 22(6), 663-670 (1990).
265. P. Balaguru, R. Narahari and M. Patel, *ACI Mater. J.* 89(6), 541-546 (1992).
266. A.S. Ezeldin and P.N. Balaguru, *J. Mater. Civil Eng.* 4(4), 415-429 (1992).
267. N.M.P. Low and J.J. Beaudoin, *Cem. Concr. Res.* 22(5), 981-989 (1992).
268. P.-W. Chen, X. Fu and D.D.L. Chung, *ACI Mater. J.* 94(2), 147-155 (1997).
269. H. Yan, W. Sun and H. Chen, *Cem. Concr. Res.* 29(3), 423-426 (1999).

Appendix A:
Electrical Behavior of Various Types of Materials

A.1 METALS

Metals can be defined as solids in which the valence electronic energy bands are not completely filled. The Fermi energy of a metal lies within the valence electronic energy band. Since the valence electrons can be excited without crossing an energy band gap, metals are characterized by high electrical conductivities. Because of overlap between adjacent valence bands, all of the valence electrons serve as charge carriers; the larger the number of valence electrons per atom, the higher the carrier concentration in the solid. For example, Al has three valence electrons per atom with the three s and three p bands overlapping, so it has a higher carrier concentration than Na, which has one valence electron per atom. Note that this definition of a metal agrees with the fact that atoms in a metal are held together by metallic bonding, which is provided by delocalized valence electrons. The delocalization of the valence electrons is necessary for them to move large distances from the parent nuclei.

A.2 INSULATORS

Whereas metals have incompletely filled valence bands, those of insulators are completely filled. Furthermore, the energy band gap between the top of the filled valence band and the bottom of the empty higher energy band is large, typically greater than 4 eV. These energy bands are illustrated in Figure A.1. In order for the valence electrons to be excited, their energy must increase enough to cross the energy band gap and move to the bottom of the empty energy band that is higher in energy. After making this transition, the electron can increase its kinetic energy and be able to move in response to an electric field. Therefore, the energy band above the valence band is known as the "conduction band" and the electrons in the conduction band are known as "conduction electrons," which serve as charge carriers in electrical conduction. In some insulators (e.g., diamond), the energy band gap is so large that it is unlikely for thermal energy to be sufficient to cause any of the valence electrons to move to the conduction band. The concentration of conduction electrons is negligible in such an insulator. In some insulators, the low concentration of conduction electrons is due to the presence of carrier traps, such as defects. Because the

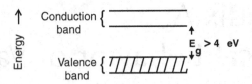

FIGURE A.1 Energy bands of an insulator. Shaded energy ranges are occupied by electrons.

carrier concentration is low, such a solid exhibits a low electrical conductivity; that is, it is an insulator.

A.3 COMPOSITES

A composite is a material formed by the artificial blending of two or more components. For example, one component is metal particles while the other is a polymer such that the metal particles are dispersed in the polymer matrix. As polymers are usually insulators, a polymer-matrix composite that is electrically conducting usually contains an electrically conducting filler (particles, flakes, short fibers, or continuous fibers).

The conductivity of a composite with an insulating matrix depends on the conductivity, volume fraction, unit size, and aspect ratio of the filler. It also depends on the distribution of the filler in the composite. The distribution, in turn, depends on the composite fabrication method. In general, the conductivity of the composite does not increase linearly with the filler volume fraction. It increases abruptly at a certain filler volume fraction called the "percolation threshold." This threshold is the filler volume fraction at which the filler units begin to touch one another sufficiently so that somewhat continuous electrical conduction paths form in the composite. Below and above the threshold the conductivity still increases with the filler volume fraction, but more gradually. A large aspect ratio of the filler enhances the touching of adjacent filler units, thus decreasing the percolation threshold. In the case of a composite formed from a mixture of insulator matrix powder and filler units, the percolation threshold is lower when the ratio of the filler unit size to the matrix particle size is smaller, as the small filler units line up along the boundaries of the large matrix particles.

The conductivity of a composite with an insulating matrix depends not only on the filler, but also on the filler-matrix interface. A stronger interface is associated with a lower contact resistivity, resulting in a lower volume resistivity for the composite. Below the percolation threshold, the filler-matrix interface plays a particularly important role. The contact resistivity (ρ_c) of an interface is a quantity that does not depend on the area of the interface or contact, but only on the nature of the interface. The contact resistance (R_c) decreases with increasing contact area (A), i.e.,

$$R_c \propto \frac{1}{A}$$

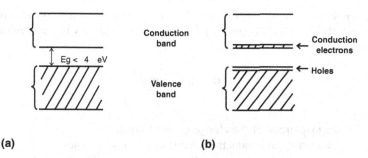

FIGURE A.2 Energy bands of a semiconductor. (a) At very low temperatures — negligible number of conduction electrons. (b) At higher temperatures — some conduction electrons. Shaded energy ranges are occupied by electrons.

The proportionality constant is ρ_c, so that

$$R_c = \frac{\rho_c}{A} \tag{A.1}$$

Since the unit of R_c is Ω and that of A is m^2, the unit of ρ_c is $\Omega.m^2$.

A.4 SEMICONDUCTORS

Semiconductors, like classical insulators, have completely filled valence bands. However, the energy band gap between the top of the filled valence band and the bottom of the conduction band is small, typically less than 4 eV. As a result, thermal energy causes a small fraction of the valence electrons to move across the energy band gap into the bottom of the conduction band, giving rise to a small number of conduction electrons that serve as charge carriers for electrical conductions. This is illustrated in Figure A.2

The transition of a small fraction of the valence electrons from the top of the valence band to the bottom of the conduction band also gives rise to unoccupied electron states at the top of the valence band. This is analogous to the vacancies in a crystal. The movement of atoms can be described by the movement of vacancies in the opposite direction. Since the number of vacancies is a small fraction of the number of atom sites in the crystal, it is simpler to describe atomic relocation by the movement of the vacancies than by the movement of the atoms themselves. A similar situation applies to the electronic vacancies, called "holes," in the top of the valence band of a semiconductor. Since the number of holes is much smaller than the number of electrons in the valence band, the transitions in energy of the valence electrons within the valence band (which is not completely filled, as shown in Figure A.2(b)) can be described more simply by the transitions in energy of the holes within the valence band. The holes provide an additional source of charge carriers. Since holes are due to the absence of electrons, they are positively charged. Hence, the charge of a hole is equal in magnitude and opposite in sign to that of an electron.

A semiconductor has two types of charge carriers, conduction electrons and holes; and, the electrical conductivity of a semiconductor has two contributions, as given by

$$\sigma = qn\mu_n + qp\mu_p \qquad (A.2)$$

where

q = magnitude of the charge of an electron
n = number of conduction electrons per unit volume
p = number of holes per unit volume
μ_n = mobility of conduction electrons
μ_p = mobility of conduction holes

The subscript "n" is used for electrons, which are negatively charged; the subscript "p" is used for holes, which are positively charged. Since the transition of each electron from the valence band to the conduction band simultaneously gives rise to a conduction electron in the conduction band and a hole in the valence band, the number of conduction electrons must equal the number of holes, i.e., n = p. Therefore, for a semiconductor (pure), Eq. (A.2) can be simplified as

$$\sigma = qn(\mu_n + \mu_p) \qquad (A.3)$$

In the presence of both a voltage gradient and a charge carrier concentration gradient, the current density in a semiconductor has a contribution from the conduction electrons and another contribution from the holes. The contribution by the conduction electrons is given by

$$\tilde{J}_n = qn\mu_n\Sigma + qD_n\frac{dn}{dx} \qquad (A.4)$$

The contribution by the holes is given by

$$\tilde{J}_p = qn\mu_p\Sigma - qD_p\frac{dnp}{dx} \qquad (A.5)$$

The total current density $\tilde{J}$ is given by

$$\tilde{J} = \tilde{J}_n + \tilde{J}_p \qquad (A.6)$$

The energy band gap is related to the ionization energy of the atoms in the solid. This is because the transition of an electron from the valence band to the conduction band means delocalizing and freeing the electron from its parent atom, thereby ionizing the parent atom. The higher the ionization energy, the larger the energy band gap. For example, consider the elements in group IV A of the periodic table.

TABLE A.1
Energy Band Gaps and Electrical Conductivities of the Elements in Group IVA of the Periodic Table

Element	Energy Band Gap (eV)	Electrical Conductivity at 20°C ($\Omega^{-1}.cm^{-1}$)
C (diamond)	~6	10^{-18}
Si	1.1	5×10^{-6}
Ge	0.72	0.02
Sn (gray)	0.08	10^4

The ionization energies of these elements increase up the group, so the energy band gaps of these elements also increase up the group, as shown in Table A.1. C (diamond), Si, Ge, and Sn (gray) all exhibit the diamond structure. They all have completely filled valence bands that contain sp^3 hybridized electrons. The difference in electrical conductivity among these elements is due to the difference in energy band gap. The gap is particularly large in diamond, so diamond is an insulator. The other three substances are all semiconductors. The smaller the energy band gap, the greater the number of valence electrons that have sufficient thermal energy to cross the energy band gap, and the higher the electrical conductivity.

Semiconductors have electrical conductivities that are between those of metals and insulators. Other than elemental semiconductors (e.g., Si, Ge, etc.), there are compound semiconductors (ZnS, GaP, GaAs, InP, etc.). In contrast to many elemental semiconductors, which exhibit the diamond structure, most compound semiconductors exhibit the zinc blende structure.

Appendix B: Temperature Dependence of Electrical Resistivity

B.1 METALS

A metal has only one type of charge carrier — electrons. The electrical conductivity of a metal is given by Eq. (B.1):

$$\sigma = qn\mu \qquad (B.1)$$

where

q	= magnitude of the charge of an electron
n	= number of free electrons (valence electrons) per unit volume
μ	= mobility of electrons

The number of valence electrons does not change with temperature, so n is independent of temperature. On the other hand, as the temperature increases, the mobility of the electrons decreases. This is because thermal vibrations of the atoms in the solid increase in amplitude as the temperature increases, and such vibrations interfere with the motion of the electrons, thus decreasing their mobility. Since n is fixed and μ decreases slightly with increasing temperature, the electrical conductivity of a metal decreases slightly with increasing temperature. In other words, the electrical resistivity of a metal increases with increasing temperature. The variation of the electrical resistivity with temperature is roughly linear, as shown schematically in Figure B.1. The rate of increase of electrical resistivity with temperature is commonly described by the temperature coefficient of electrical resistivity (α), which is defined by the equation

$$\frac{\Delta\rho}{\rho} = \alpha\Delta T \qquad (B.2)$$

where $\Delta\rho/\rho$ is the fractional change in electrical resistivity, and ΔT is the increase in temperature. The value of α is ~0.004°C^{-1} for most pure metals. Note that Eq. (B.2) requires that the unit of α be °C^{-1}, which is the same as K^{-1}, where K means "Kelvin" or absolute temperature.

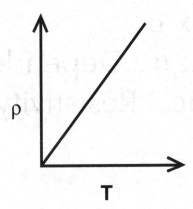

FIGURE B.1 The temperature dependence of the electrical resistivity of a metal.

B.2 SEMICONDUCTORS

A semiconductor has two types of carriers — conduction electrons and holes. The electrons (negative) move up a voltage gradient, whereas the holes (positive) move down a voltage gradient, as illustrated in Figure B.1. Conventionally, the direction of the current is taken as the direction of the flow of the positive charge carriers, i.e., from a higher voltage to a lower voltage.

The electrical conductivity of an intrinsic semiconductor is given by Eq. (A.3), which is

$$\sigma = qn(\mu_n + \mu_p) \tag{A.3}$$

Due to thermal vibrations of the atoms, both μ_n and μ_p decrease slightly as the temperature increases. Since thermal energy causes the valence electrons to be excited to the conduction band, the value of n increases with increasing temperature. As for any thermally-activated process, the temperature dependence is exponential. It can be shown that n varies with temperature in the form

$$n \propto e^{-E_g/2kT}$$

where
  E_g = energy band gap between conduction and valence bands
  k   = Boltzmann's constant
  T   = temperature in K

The factor of 2 in the exponent is because the excitation of an electron across E_g produces an intrinsic conduction electron and an intrinsic hole.

Since μ_n and μ_p only decrease slightly with increasing temperature, whereas n significantly increases with increasing temperature, the variation of σ with temperature is roughly of the form,

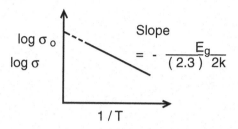

FIGURE B.2 An Arrhenius (semi-logarithmic) plot of the variation of the electrical conductivity with temperature for an intrinsic semiconductor.

$$\sigma \propto e^{-E_g/2kT}$$

E_g is itself temperature-dependent, although its slight temperature dependence is negligible in most circumstances. Let the constant of proportionality be σ_0. Taking natural logarithms,

$$\sigma = \sigma_0 e^{-E_g/2kT}$$

$$\ln\sigma = \ln\sigma_0 - \frac{E_g}{2kT} \tag{B.3}$$

Changing the natural logarithms to logarithms of base 10,

$$\log\sigma = \log\sigma_0 - \frac{E_g}{(2.3)2kT} \tag{B.4}$$

Thus, the variation of σ with T can be shown in an Arrhenius plot of log σ against $1/T$, as shown in Figure B.2. The plot is a straight line of slope $-[E_g/(2.3)2k]$; the intercept of the line on the log σ axis is log σ_0. Therefore, by measuring σ as a function of T, one can determine E_g and σ_0. This is one of the most common ways to determine the energy band gap of a semiconductor.

The significant variation of σ with T for a semiconductor allows a semiconductor to serve a temperature measuring device called a "thermistor." The temperature is indicated by the conductivity, which relates to the measured resistance of the thermistor. The resistance decreases with increasing temperature.

The most clearcut difference in behavior between a metal and a semiconductor is that the electrical conductivity of a metal decreases with increasing temperature, whereas that of a semiconductor increases with increasing temperature. Therefore, the simplest way to determine whether a solid is a metal or a semiconductor is to see whether the electrical conductivity of the solid increases or decreases with temperature.

Appendix C:
Electrical Measurement

The most basic form of electrical measurement involves measurement of the resistance (DC) or impedance (AC). The impedance Z is a complex quantity having the resistance R as the real part and the reactance X as the imaginary part, i.e.,

$$Z = R + jX \tag{C.1}$$

where j represents $\sqrt{-1}$. The reactance X is due to capacitance and/or inductance. The quantities Z, R, and X depend on frequency υ (angular frequency $\omega = 2\pi\upsilon$)
For a capacitance C,

$$X = -\frac{1}{\omega C} \tag{C.2}$$

Note that X, in this case, decreases with increasing ω. The current through a capacitor leads the voltage by a phase of 90°, because the current is

$$i = \frac{dq}{dt} \tag{C.3}$$

where q is the charge and

$$q = Cv \tag{C.4}$$

where v is the voltage (AC), i.e., $v = V \sin \omega t$.
For an inductance L,

$$X = \omega L \tag{C.5}$$

Note that X in this case increases with increasing ω. The current i through an inductor lags the voltage v by a phase of 90° because

$$v = L\frac{di}{dt} \tag{C.6}$$

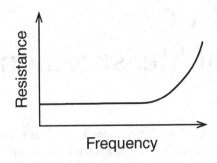

FIGURE C.1 Variation of resistance R with frequency due to the skin effect.

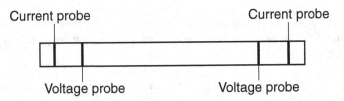

FIGURE C.2 The four-probe method of resistance/impedance measurement.

An electric field at a high frequency penetrates only the near surface region of a conductor. The amplitude of the wave decreases exponentially as the wave penetrates the conductor. The depth at which the amplitude is decreased to 1/e of the value at the surface is called the "skin depth," and the phenomenon is known as the "skin effect." The higher the frequency υ, the smaller the skin depth δ, which is given by

$$\delta = (\sigma\pi\mu\upsilon)^{-1/2} \tag{C.7}$$

where σ is the DC electrical conductivity, μ is the permeability ($\mu = \mu_0\mu_r$), and υ is the frequency. The greater is σ, the smaller is δ. Because of the skin effect, only the surface region contributes significantly to conduction. Hence, R increases at high frequencies, as illustrated in Figure C.1.

Measurement of resistance (DC) or impedance (Z) typically involves a sample that has its long dimension along the direction of resistance/impedance measurement, since the resistance/impedance increases with length. This measurement is best conducted by using the four-probe method, which involves four electrical contacts such that the outer two are for passing current and the inner two for measuring voltage, as illustrated in Figure C.2. Voltage divided by current is resistance (DC) or impedance (AC). In contrast, the two-probe method uses two electrical contacts, such that each contact is for both passing current and measuring voltage, as illustrated in Figure C.3. The main advantage of the four-probe over the two-probe method is that the contact resistance/impedance is not included in the measured resistance/impedance in the former, but is included in the latter. Since the contact resis-

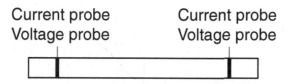

FIGURE C.3 The two-probe method of resistance/impedance measurement.

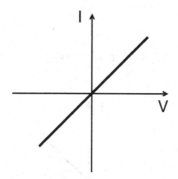

FIGURE C.4 I-V characteristic for ohmic behavior.

tance/impedance may be even larger than the sample resistance/impedance, excluding it is important. The contact resistance/impedance is not included in the four-probe method because no current passes through the voltage probes and, consequently, there is no contact potential drop (i.e., $V = IR = 0$ when $I = 0$ and $R \neq 0$). Another advantage of the four-probe method is that the current path is parallel to the direction of resistance/impedance measurement in the part of the sample between the voltage probes. In contrast, the current path near the two probes is not parallel in the two-probe method. Having parallel current paths means that a cross-sectional plane perpendicular to the current paths is equipotential.

The electrical contacts may be made by using solder, electrically conducting adhesive, electrically conducting paint, or pressure. Solder suffers from the requirement of heating during application, and the heating may not be acceptable to the material. An adhesive or paint does not require heating, although some adhesives perform better if heated. A pressure contact is the least expensive and disconnection is easy, but it suffers from high contact resistivity.

Resistance/impedance measurements may be performed at different positive and negative voltage levels. If the resulting plot of current vs. voltage (called the "I-V characteristic") is a straight line through the origin (Figure C.4), the resistance/impedance is independent of the voltage and the behavior is said to be ohmic. If the current is zero when the voltage is negative, and non-zero when the voltage is positive (Figure C.5), the behavior is said to be rectifying. The I-V characteristic provides information on the electrical behavior. When the voltage is very high, breakdown (as in dielectric breakdown, Appendix D) can occur, causing the I-V characteristic to deviate from linearity (Figure C.6).

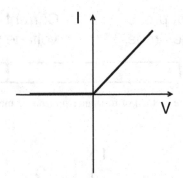

FIGURE C.5 I-V characteristic for rectifying behavior.

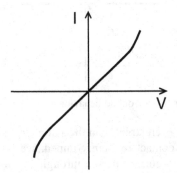

FIGURE C.6 I-V characteristic for ohmic behavior with dielectric breakdown.

Measurement of the resistance/impedance does not give information on the type of carrier, as holes and electrons both respond to the applied voltage and yield currents in the same direction. To obtain information on the type of carrier, either Hall measurement or thermopower measurement can be conducted.

Hall measurement involves measuring the voltage (called "Hall voltage") in the vertical direction in the plane of Figure C.7 while the carriers move in the horizontal direction in response to the applied voltage V, and the applied magnetic field is in the direction going into the page. The carriers (each of charge q) experience a force $\vec{F}$ (called the "Lorentz force") in the vertical direction, such that

$$\vec{F} = q\vec{v}X\vec{B} \tag{C.8}$$

where $\vec{v}$ is the velocity of the carriers and B is the magnetic induction. If q is positive (i.e., the carriers are holes, which move toward the right in response to the applied voltage V), $\vec{F}$ is in the upward direction. This force causes the holes to move upward. Thus, the Hall voltage V_H is positive. If q is negative (i.e., the carriers are electrons,

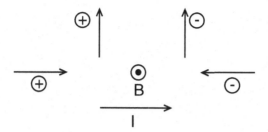

FIGURE C.7 Hall measurement.

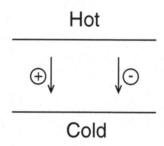

FIGURE C.8 Thermopower measurement.

which move toward the left in response to the applied voltage V), $\vec{F}$ is also in the upward direction. This force causes the electrons to move upward. Thus, the Hall voltage V_H is negative. From the sign of V_H, the type of carrier can be identified.

Thermopower measurement involves applying a temperature gradient across a sample. The carriers have a net flow from the hot end to the cold end because the greater thermal energy at the hot end causes more carrier motion in the vicinity of the hot end. If the carriers are holes, their movement will cause the cold end to be at a positive voltage relative to the hot end (Figure C.8). On the other hand, if the carriers are electrons, their movement will cause the hot end to be at a positive voltage relative to the cold end. Thus, from the sign of the voltage, the type of the carrier can be identified. The "thermopower" is defined as the voltage per unit temperature difference.

Appendix D: Dielectric Behavior

Electrical insulators are also known as dielectrics. Most ionic solids and molecular solids are insulators because of the negligible concentration of conduction electrons or holes.

Consider two metal (conductor) plates connected to the two ends of a battery of voltage V (Figure D.1). Assuming the electrical resistance of the connecting wires is negligible, the potential between the two metal plates is V. Assume that the medium between the plates is a vacuum.

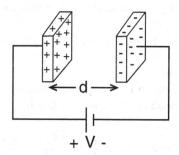

FIGURE D.1 A pair of positively charged and negatively charged conductor plates in vacuum.

There are a lot of conduction electrons in the metal plates and metal wires. The positive end of the battery attracts conduction electrons, making the left plate positively charged (charge = +Q). The negative end of the battery repels conduction electrons, making the right plate negatively charged (charge = –Q). Let the area of each plate be A. the magnitude of charge per unit area of each plate is known as the "charge density" (D_o).

$$D_o = \frac{Q}{A} \tag{D.1}$$

The electric field E between the plates is given by

$$E = \frac{V}{d} \tag{D.2}$$

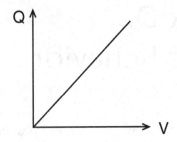

FIGURE D.2 Plot of charge Q vs. potential V. The slope $= C_o = \dfrac{\varepsilon_o A}{d}$.

where d is the separation of the plates. When $E = 0$, $D_o = 0$. In fact, D is proportional to E. Let the proportionality constant be ε_o. Then

$$D_o = \varepsilon_o E \tag{D.3}$$

ε_o is called the "permittivity of free space." It is a universal constant.

$$\varepsilon_o = 8.85 \times 10^{-12} \text{ C/(V.m)}$$

Eq. (D.3) is known as Gauss's law.

Just as D_o is proportional to E, Q is proportional to V. The plot of Q versus V is a straight line through the origin (Figure D.2), with

$$\text{Slope} = C_o = \frac{Q}{V} = \frac{\varepsilon_o E A}{Ed} = \frac{\varepsilon_o A}{d} \tag{D.4}$$

C_o is known as the "capacitance." Its unit is Coulomb/Volt, or Farad (F). In fact, this is the principle behind a parallel-plate capacitor.

Next, consider that the medium between the two plates is not a vacuum, but an insulator whose center of positive charge and center of negative charge coincide when $V = 0$. When $V > 0$, the center of positive charge is shifted toward the negative plate, while the center of negative charge is shifted toward the positive plate. Such displacement of the centers of positive and negative charges is known as "polarization." In the case of a molecular solid with polarized molecules (e.g., HF), the polarization in the molecular solid is due to the preferred orientation of each molecule such that the positive end of the molecule is closer to the negative plate. In the case of an ionic solid, the polarization is due to the slight movement of the cations toward the negative plate, and that of the anions toward the positive plate. In the case of an atomic solid, the polarization is due to the skewing of the electron clouds toward the positive plate.

When polarization occurs, the center of positive charge sucks more electrons to the negative plate, causing the charge on the negative plate to be $-\kappa Q$, where $\kappa > 1$ (Figure D.3). Similarly, the center of negative charge repels more electrons away

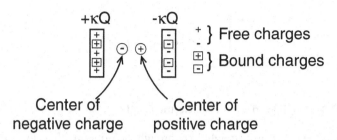

FIGURE D.3 A pair of positively charged and negatively charged conductor plates in a medium with relative dielectric constant κ.

from the positive plate, causing the charge on the positive plate to be κQ. κ is a unitless number called the "relative dielectric constant." Its value at 1 MHz (10^6 Hz) is 2.3 for polyethylene, 3.2 for polyvinyl chloride, 6.5 for Al_2O_3, 3000 for $BaTiO_3$, and 78.3 for water.

The charges in the plates when a vacuum is between the plates are called "free charges" (magnitude = Q on each plate). The extra charges in the plates when an insulator is between the plates are called the "bound charges" (magnitude = $\kappa Q - Q = (\kappa - 1)Q$ on each plate).

When an insulator is between the plates, the charge density is given by

$$D_m = \kappa D_o = \frac{\kappa Q}{A} \tag{D.5}$$

Using Eq. (D.3), Eq. (D.5) becomes

$$D_m = \kappa \varepsilon_o E = \varepsilon E \tag{D.6}$$

where $\varepsilon \equiv \kappa \varepsilon_o$; ε is known as the dielectric constant, whereas κ is known as the relative dielectric constant. Hence, ε and ε_o have the same unit.

When an insulator is between the plates, the capacitance is given by

$$C_m = \frac{\kappa Q}{V} = \frac{\kappa \varepsilon_o E A}{Ed} = \frac{\kappa \varepsilon_o A}{d} = \kappa C_o \tag{D.7}$$

From Eq. (D.7), the capacitance is inversely proportional to d, so capacitance measurement provides a way to detect changes in d (i.e., to sense strain).

Mathematically, the polarization is defined as the bound charge density, so that it is given by

$$\begin{aligned} P &= D_m - D_o \\ &= \kappa \varepsilon_o E - \varepsilon_o E \\ &= (\kappa - 1)\,\varepsilon_o E \end{aligned} \tag{D.8}$$

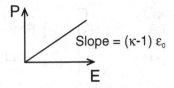

FIGURE D.4 Plot of polarization P vs. electric field E.

The plot of P versus E is a straight line through the origin, with a slope of $(\kappa - 1)\,\varepsilon_0$ (Figure D.4).

The ratio of the bound charge density to the free charge density is given by

$$\frac{\kappa Q - Q}{Q} = \kappa - 1 \qquad (D.9)$$

The quantity $\kappa - 1$ is known as the "electric susceptibility" (χ). Hence,

$$\chi = \kappa - 1 = \frac{P}{\varepsilon_0 E} \qquad (D.10)$$

The dipole moment in the polarized insulator is given by

$$(\text{bound charge})d = (\kappa - 1)\,Qd \qquad (D.11)$$

since the bound charges are induced by the dipole moment in the polarized insulator. The dipole moment per unit volume of the polarized insulator is thus given by

$$\frac{\text{Dipole moment}}{\text{Volume}} = \frac{(\kappa - 1)Qd}{Ad} = \frac{(\kappa - 1)Q}{A} = P \qquad (D.12)$$

Therefore, another meaning of polarization is the dipole moment per unit volume.

Now consider that the applied voltage V is an AC voltage, so that it alternates between positive and negative values. When $V > 0$, $E > 0$, and the polarization is one way. When $V < 0$, $E < 0$, and the polarization is in the opposite direction. If the frequency of V is beyond about 10^{10} Hz, the molecules in the insulator cannot reorient themselves fast enough to respond to V (i.e., dipole friction occurs), so κ decreases (Figure D.5). When the frequency is beyond 10^{15} Hz, even the electron clouds cannot change their skewing directions fast enough to respond to V, so κ decreases further. The minimum value of κ is 1, which is the value for vacuum.

In making capacitors, one prefers to use insulators with very large values of κ so that C_m is large. However, one should be aware that the value of κ depends on the frequency. It is challenging to make a capacitor that operates at very high frequencies.

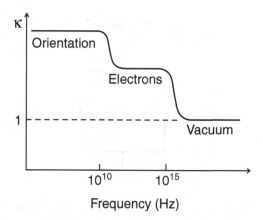

FIGURE D.5 Dependence of the relative dielectric constant κ on the frequency.

In AC condition, it is mathematically more convenient to express E and D_m in complex notation, i.e.,

$$E = \hat{E}e^{i\omega t} = \hat{E}(\cos \omega t + i \sin \omega t) \tag{D.13}$$

and

$$D_m = \hat{D}_m e^{i(\omega t - \delta)} = \hat{D}_m[\cos(\omega t - \delta) + i\sin(\omega t - \delta)] \tag{D.14}$$

where ω is the angular frequency (which is related to the frequency f by the equation $\omega = 2\pi f$; angular frequency has the unit radians/s, whereas frequency has the unit cycle/s, or Hertz (Hz)), δ is the dielectric loss angle (which describes the lag of D_m with respect to E), $\hat{E}$ is the amplitude of the E wave, and $\hat{D}_m$ is the amplitude of the D_m wave. If $\delta = 0$, E and D_m are in phase and there is no lag. Not only are E and D_m complex, ε is complex too. Eq. (D.6) then becomes

$$\hat{D}_m e^{i(\omega t - \delta)} = \varepsilon \hat{E} e^{i\omega t}$$

Hence,

$$\varepsilon = \frac{\hat{D}_m}{\hat{E}} e^{-i\delta} = \frac{\hat{D}_m}{\hat{E}}(\cos\delta - i\sin\delta) \tag{D.15}$$

The quantity $\hat{D}_m/\hat{E}$ is known as the "static dielectric constant." From Eq. (D.15), the loss factor (or loss tangent or dielectric loss or dissipation factor) tan δ is given by

$$\tan\delta = -\frac{\text{Imaginary part of } \varepsilon}{\text{Real part of } \varepsilon} \tag{D.16}$$

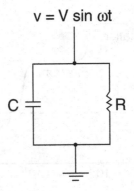

FIGURE D.6 A nonideal capacitor modeled as a capacitor C and a resistor R in parallel.

In another convention, the dissipation factor is defined as $\tan \delta$, while the dielectric loss factor is defined as $\kappa \tan \delta$.

By definition, $\varepsilon = \kappa \varepsilon_o$, so Eq. (D.16) can be written as

$$\tan \delta = -\frac{\text{Imaginary part of } \kappa}{\text{Real part of } \kappa} \qquad \text{(D.17)}$$

The loss factor $\tan \delta$ is more commonly used than δ itself to describe the extent of lag between D_m and E. This is because $\tan \delta$ relates to the energy loss. The value of δ at 1 MHz (10^6 Hz) is 0.0001 for polyethylene, 0.05 for polyvinyl chloride, and 0.001 for Al_2O_3.

Lag occurs when the insulator is not a perfect insulator. A nonideal capacitor can be modeled as a capacitor C in parallel with a resistor (DC) R, as shown in Figure D.6. If the insulator is perfect, R will be infinite and thus disappear from Figure D.6. The current i_c through the capacitor is given by

$$i_c = \frac{dQ}{dt} = C\frac{dv}{dt} \qquad \text{(D.18)}$$

where v is the AC voltage across both capacitor and resistor. Let

$$v = V \sin \omega t \qquad \text{(D.19)}$$

where V is the amplitude of the voltage wave of angular frequency ω, which is related to the frequency f and period T by

$$\omega = 2\pi f = \frac{2\pi}{t} \qquad \text{(D.20)}$$

Substituting Eq. (D.19) into Eq. (D.18),

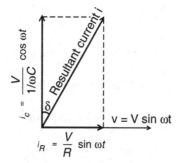

FIGURE D.7 Phasor diagram for the nonideal capacitor of Figure D.6.

$$i_c = C\frac{dv}{dt} = \omega CV \cos\omega t \qquad (D.21)$$

$$= \frac{V}{1/(\omega C)}\cos\omega t$$

Since

$$\sin\left(\omega t + \frac{\pi}{2}\right) = \sin\omega t\cos\frac{\pi}{2} + \cos\omega t\sin\frac{\pi}{2}$$

$$= \cos\omega t$$

Eq. (D.21) becomes

$$i_c = \frac{V}{1/(\omega C)}\sin\left(\omega t + \frac{\pi}{2}\right) \qquad (D.22)$$

Comparison of Eq. (D.22) with Eq. (D.19) shows that i_c leads v by a phase angle of $\pi/2$, or 90°. This pertains to the current through the capacitor only.

The current through the resistor R is

$$i_R = \frac{v}{R} = \frac{V}{R}\sin\omega t \qquad (D.23)$$

i_R and v are in phase.

The phase relationship is conventionally described by a phasor diagram (Figure D.7), which shows that i_R is in phase with v, i_C is 90° ahead of v, and the resultant current i is ahead of v by $\phi = 90° - \delta$. From Figure D.7,

$$\tan\delta = \frac{V/R}{V\omega C} = \frac{1}{\omega CR} \qquad (D.24)$$

The electrical energy stored in the perfect capacitor is

$$\text{Energy stored} = \int_0^\tau v i_C \, dt \tag{D.25}$$

$$= \int_0^\tau V^2 \omega C \sin \omega t \cos \omega t \, dt$$

$$= \int_0^\tau \frac{V^2 \omega C}{2} \sin 2\omega t \, dt$$

$$= -\frac{V^2 \omega C}{4\omega(2)} [\cos 2\omega]_0^\tau$$

$$= -\frac{1}{4} CV^2 (\cos 2\omega t - 1)$$

The maximum value of this energy is $\frac{1}{2} CV^2$, and this occurs when $\cos 2\omega t = -1$.

The higher is C, the greater the energy loss.

The loss of energy per cycle due to conduction through the resistor R is

$$\text{Energy loss} = \frac{V^2}{R} \int_0^{2\pi/\omega} \sin \omega t \sin \omega t \, dt \tag{D.26}$$

$$= \frac{V^2}{\omega R} \int_0^{2\pi} \frac{1}{2}(1 - \cos 2\omega t) d(\omega t)$$

$$= \frac{V^2}{\omega R} \left[\frac{1}{2}\left(\omega t - \frac{1}{2}\sin 2\omega t \right) \right]_0^{2\pi}$$

$$= \frac{V^2}{\omega R} \left[\frac{1}{2}(2\pi - 0 - 0 + 0) \right]$$

$$= \frac{V^2 \pi}{\omega R}$$

The smaller is R, the greater is the energy loss. When $R = \infty$, R becomes an open circuit and disappears from the circuit of Figure D.6.

$$\frac{\text{Energy lost per cycle}}{2\pi \times \text{maximum energy stored}} = \frac{V^2 \pi / \omega R}{2\pi CV^2 / 2} \tag{D.27}$$

$$= \frac{1}{\omega CR} = \tan \delta$$

Hence, $\tan \delta$ is related to the energy loss. Energy loss is undesirable for most piezoelectric applications.

Dipole friction causes energy loss such that the greatest loss occurs at frequencies at which dipole orientation can almost, but not quite, occur. This is shown in Figure D.8, which corresponds to Figure D.5. A peak in $\tan \delta$ occurs at each step decrease in κ as the frequency increases.

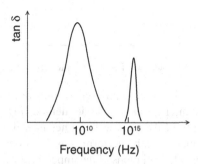

FIGURE D.8 Dependence of tan δ on the frequency.

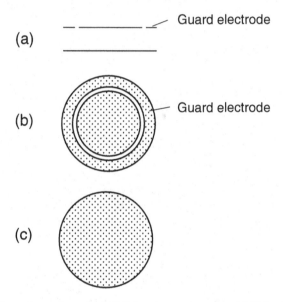

FIGURE D.9 Three-electrode method of dielectric measurement: (a) Side view, (b) top view, (c) bottom view.

Dielectric measurement refers to measurement of the relative dielectric constant κ of a dielectric material. Parallel plate geometry is used because the capacitance increases with plate area. This geometry involves sandwiching the sample between electrical conductor plates, applying a voltage across the parallel plates, and measuring the impedance across the plates. The measurement is best conducted using the three-electrode method, which involves one of the plates being one electrode, and the other plate consisting of an outer electrode and an inner electrode that are electrically unconnected (Figure D.9). The outer electrode, called the "guard electrode," ensures that the electric flux lines across the inner electrode and the other plate are all perpendicular to the plates so that a cross-section of the sample parallel to the plates is equipotential. On the other hand, the electric flux lines across the

FIGURE D.10 Two-electrode method of dielectric measurement, side view.

outer electrode and the other plate partly curve outward. In contrast to the three-electrode method is the inferior two-electrode method (Figure D.10), which suffers from the outward curving of the electric flux lines near the edge of the sample.

Dielectric measurements can be performed as functions of frequency as well as voltage across the plates. The variations of the impedance with frequency and voltage provide information on the origin of dielectric behavior. The temperature may also be varied to yield even more information.

Appendix E:
Electromagnetic Measurement

An electromagnetic measurement involves an electromagnetic wave that has in it both electric and magnetic fields that are perpendicular to the direction of propagation of the wave (a transverse wave). A common technique to guide an electromagnetic wave is to use a coaxial cable, a cable with a circular cross-section such that the circular metal wall serves as the electrical ground while an axial metal pin at the center of the circle serves as the signal-carrying conductor (Figure E.1). The electric field is radial in the plane of the circle as the wave propagates along the axis of the cable.

By placing a circular sample (with a hole at its center for the axial conductor) across the whole cross-section of a fixture resembling a coaxial cable (except larger), the interaction of the radiation with the sample can be studied. This method is called the "coaxial cable method" (also known as the "transfer impedance method"). Electromagnetic radiation incident on a sample may be partly reflected, partly absorbed, and partly transmitted. Measurement of the intensities of the transmitted wave and reflected wave can be achieved by using a network analyzer, an electronic instrument with source, receiver, and frequency sweep capability. The smaller the intensity of the transmitted wave, the greater the EMI shielding effectiveness.

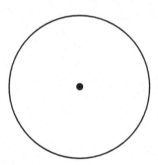

FIGURE E.1 A coaxial cable.

Appendix F: Thermoelectric Behavior

Thermoelectric phenomena involve the transfer of energy between electric power and thermal gradients. They are widely used for cooling and heating, including air conditioning, refrigeration, thermal management, and the generation of electrical power from waste heat.

The thermoelectric phenomenon involving the conversion of thermal energy to electrical energy is embodied in the Seebeck effect, i.e., the greater concentration of carrier above the Fermi energy at the hot point than at the cold point, the consequent movement of mobile carrier from the hot point to the cold point, and the resulting voltage difference (called the "Seebeck voltage") between the hot and cold points. If the mobile carrier is electrons, the hot point is negative in voltage relative to the cold point (Figure F.1(a)). If the mobile carrier is holes, the cold point is positive relative to the hot point (Figure F.1(b)). Hence, a temperature gradient results in a voltage. The change in Seebeck voltage per degree C temperature rise is called the "thermoelectric power," the "thermopower," or the "Seebeck coefficient," which is positive for the polarity illustrated in Figure F.1(b).

A thermocouple is made up of two dissimilar materials that are joined at one end to form a junction (Figure F.2). When the two materials are subjected to the same temperature gradient, the Seebeck voltage (V_1 in Material 1, and V_2 in Material 2) developed in the two materials are different since the two materials differ in the Seebeck coefficient. The voltage difference between the ends of the two materials away from the junction is the difference between V_1 and V_2. The greater the temperature gradient, the greater are V_1, V_2, and the difference between them. By measuring the voltage between the ends away from the junctions, the temperature difference between the hot and the cold ends can be obtained. In this way, the

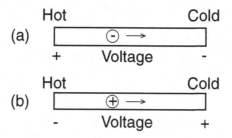

FIGURE F.1 Seebeck effect due to (a) electrons, (b) holes.

Temperature B

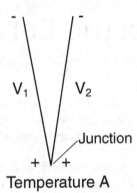

FIGURE F.2 A thermocouple comprising two dissimilar materials, which give Seebeck voltages V_1 and V_2.

thermocouple is a temperature measurement device. In using a thermocouple, the junction is placed at the location where temperature measurement is desired. The ends of the materials away from the junction are kept at a known temperature, such as room temperature. The change in voltage difference between the ends away from the junction per degree C change in temperature difference between hot and cold ends is called the "thermocouple sensitivity." Note that the thermocouple is not based on an interfacial phenomenon, but on the difference in bulk properties of the dissimilar materials.

Appendix G:
Nondestructive Evaluation

G.1 INTRODUCTION

Nondestructive evaluation (NDE) or nondestructive testing (NDT) refers to the testing of a component without harming it to assess its quality (during or after manufacture) or condition (during or after use). Assessment during manufacture enables adjustment of the manufacturing conditions of either the particular component being manufactured or similar components to be manufactured in order to improve the quality. This assessment relates to smart manufacturing and quality control. Assessment during use enables the monitoring of damage induced by use so as to mitigate hazards. This relates to smart structures and structural health monitoring. The use of embedded sensors and structural materials with the inherent abilities to sense was described in Chapters 2–6. This appendix describes sensing methods that involve materials or instrumentation that are external to the structure. These are commonly used for inspection because they are nonintrusive and widely applicable, though they are limited to sensing damage such as cracks. Strain, residual stress, and subtle damage cannot be sensed. Moreover, these types of damage may not be amenable to sensing in real time.

G.2 LIQUID PENETRANT INSPECTION

Liquid penetrant inspection detects surface defects. It involves applying a penetrant to the surface to be inspected by dipping, spraying, or brushing. The penetrant is then pulled into the surface crack by capillary action (Figure G.1(b)). After allowing sufficient time for the penetrant to be drawn into surface cracks, the excess penetrant is removed (Figure G.1(c)). After this, a developer (an absorbent material capable of drawing the penetrant from the cracks) is applied so that some penetrant is extracted to the surface (Figure G.1(d)) and visual inspection is possible. In order to make the penetrant more visible, brightly colored dyes or fluorescent materials are often added. Moreover, the developer is usually chosen to provide a contrasting background. After inspection, the developer and remaining penetrant are removed by cleaning.

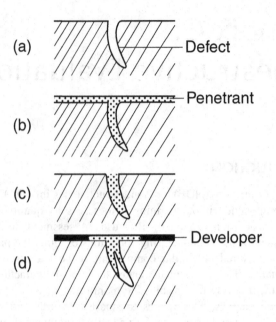

FIGURE G.1 Liquid penetrant inspection.

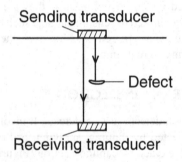

FIGURE G.2 Ultrasonic inspection involving two transducers in the through-transmission configuration.

G.3 ULTRASONIC INSPECTION

An ultrasonic wave has a higher frequency than audible sound. The frequency typically ranges from 25 to 100,000 MHz. A higher frequency means a shorter wavelength, which allows smaller defects to be detected.

Ultrasonic inspection involves sending an ultrasonic wave emitted by a pulsed oscillator and transducer through the material to be inspected and measuring the intensity of the reflected wave or the transmitted wave, as well as the time it takes for the wave to be detected. A defect such as a crack acts as a barrier for the transmission of the wave, so the intensity of the transmitted wave is decreased. This

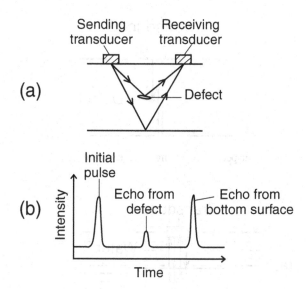

FIGURE G.3 (a) Ultrasonic inspection involving two transducers in the pulse-echo mode. (b) Plot of intensity (transducer voltage) vs. time showing the initial pulse and echoes from the bottom surface and the intervening defect.

is illustrated in Figure G.2, where two transducers are used in the through-transmission configuration.

The through-thickness (in-line) configuration also allows measurement of the time it takes for the wave to go from the sending transducer to the receiving transducer. This time, together with the distance between the two transducers, gives the ultrasonic sound speed of the material under inspection. During the curing of a thermosetting resin, the ultrasonic sound speed of the resin changes. This makes sense because the viscosity increases greatly as curing takes place. Hence, measurement of the ultrasonic sound speed allows cure monitoring. Another method of cure monitoring is the dielectric method in which the AC electrical conductivity is measured. This conductivity is the ionic conductivity, which changes during curing. Yet another method of cure monitoring uses an optical fiber.

A defect provides an interface that reflects the wave, so a reflected wave is detected at an earlier time compared to the time when the reflected wave from the bottom surface is detected. This is shown in Figure G.3, where two transducers are used in the pulse-echo mode. An oscilloscope may be used to obtain the plot in Figure G.3(b). The time of the echo from the defect gives information on its depth. The intensity of the echo gives information on its size. It is possible for the receiving and sending transducer to be the same, as shown in Figure G.4, but this is limited to the pulse-echo mode.

Since air is a poor transmitter of ultrasonic waves, an acoustic coupling medium (a liquid such as water, oil, or grease) is needed to connect a transducer to the material to be inspected. Figure G.5 illustrates the use of water as a coupling medium in a pulse-echo configuration similar to Figure G.4. Note that Figures G.5(b) and

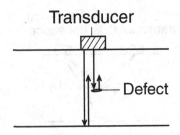

FIGURE G.4 Ultrasonic inspection involving a single transducer.

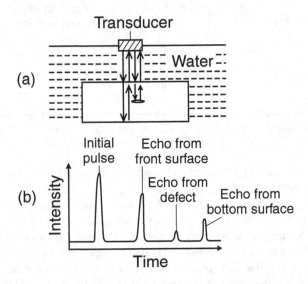

FIGURE G.5 (a) Ultrasonic inspection involving a single transducer connected to the material by water, which serves as an acoustic coupling medium. (b) Plot of intensity (transducer voltage) vs. time showing the initial pulse and echoes from the front and bottom surfaces and the intervening defect.

G.3(b) differ in that the echo from the front surface is present in the former but absent in the latter. The use of a coupling medium also means that the transducers do not have to make contact with the material under inspection. The noncontact configuration is particularly attractive when the material under inspection is hot.

An ultrasonic technique more sensitive than the technique described in Figures G.2–G.5 involves measuring the attenuation of the intensity of the ultrasonic wave upon bouncing back and forth by the front and bottom surfaces of the material under inspection. The greater the amount of defects in the material, the greater the attenuation from cycle to cycle. For example, the echo from the back surface (Figure G.4) is recorded as a function of the cycle number, as shown in Figure G.6.

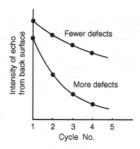

FIGURE G.6 Attenuation of ultrasonic wave upon traveling through the material. One cycle means traveling from the front surface to the back surface and then to the front surface.

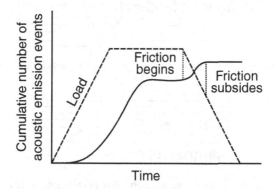

FIGURE G.7 Plot of the cumulative number of acoustic emission events vs. time (solid curve) and plot of load vs. time (dashed curve) during loading and unloading of a fiber composite experiencing delamination during loading and friction between delaminated surfaces during unloading.

G.4 ACOUSTIC EMISSION TESTING

In response to an applied stress, a material may develop defects such as cracks. The process is accompanied by the emission of ultrasonic waves. This is known as "acoustic emission" (AE). The detection of these waves provides a means of damage monitoring in real time. Detection is achieved using receiving transducers. By the use of two or more transducers in different locations of a test piece, the location of the damage can be determined. The energy of an acoustic emission event relates to the damage mechanism, e.g., fiber breakage vs. delamination in a fiber composite. Acoustic emission occurs during loading, as defects are generated during loading. However, some damage mechanisms, such as delamination, cause acoustic emission during unloading, in addition to more extensive acoustic emission during loading. In the case of delamination, this is due to frictional noise associated with the adjacent delaminated surfaces coming in contact as unloading occurs (Figure G.7).

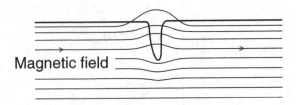

FIGURE G.8 Distortion of the magnetic flux lines due to a surface crack in a ferromagnetic or ferrimagnetic material.

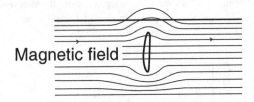

FIGURE G.9 Distortion of the magnetic flux lines due to a subsurface defect.

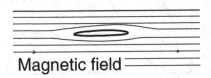

FIGURE G.10 Little distortion of the magnetic flux lines when the length of the defect is parallel to the applied magnetic field.

G.5 MAGNETIC PARTICLE INSPECTION

The magnetic flux lines in a ferromagnetic or ferrimagnetic material are distorted around a defect, since the defect (such as a crack) has a low magnetic permeability ($\mu_r = 1$ for air). This distortion for a surface crack is illustrated in Figure G.8. As a result of the distortion, magnetic flux lines protrude from the surface at the location of the surface crack. This is known as "field leakage," which attracts magnetic particles that are applied to the surface of the material to be inspected. Hence, the location of the magnetic particles indicates the location of the defect. A subsurface defect that is sufficiently close to the surface can also be detected, as shown in Figure G.9.

This method requires applying a magnetic field in a suitable direction, which is preferably perpendicular to the length of the defect, in order to maximize the distortion of the magnetic flux lines. As shown in Figure G.10, a defect that has its length parallel to the applied magnetic field does not cause much distortion of the magnetic flux lines. A magnetic field along the axis of a cylindrical sample may be applied by passing an electric current circumferentially, as shown in Figure G.11(a). A circumferential magnetic field may be applied by passing a current in the axial direction, as shown in Figure G.11(b). Depending on the direction of the magnetic field, defects of various orientations are detected. This method is restricted to ferromagnetic and ferrimagnetic materials.

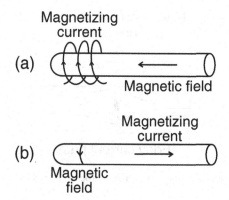

FIGURE G.11 (a) An axial magnetic field generated by a circumferential electric current. (b) A circumferential magnetic field generated by an axial electric current.

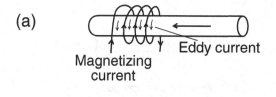

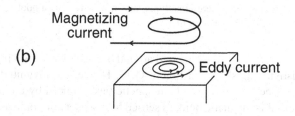

FIGURE G.12 The generation of an eddy current by an applied magnetic field: (a) a cylindrical sample, (b) a flat sample.

G.6 EDDY CURRENT TESTING

An eddy current is an electric current induced in an electrically conducting material due to an applied time-varying magnetic field. The phenomenon is due to Faraday's law, which says that a voltage is generated in a conductor loop when the magnetic flux through the loop is changed. The eddy current is in a direction such that the magnetic field it generates opposes the applied magnetic field. If a circumferential current is used to generate the applied electric field, the resulting eddy current is in the opposite direction from the magnetizing current, as shown in Figure G.12 for a cylindrical sample and a flat sample.

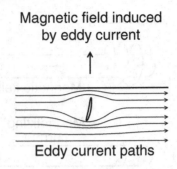

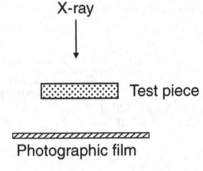

FIGURE G.13 Distortion of eddy current paths around a defect.

FIGURE G.14 X-radiography setup involving an x-ray source and a photographic film, with the material to be inspected between them.

The eddy current paths are distorted around a defect, which is a region of higher electrical resistivity, as shown in Figure G.13. Hence, the magnitude of the eddy current is diminished. As a result, the magnetic field induced by the eddy current is decreased. Thus, this magnetic field, as sensed by a sensor coil, indicated the severity of the defect. By moving the sensor coil to different locations on the sample surface, the location of the defect is detected. This method is limited to electrically conducting materials.

G.7 X-RADIOGRAPHY

Internal defects are usually best detected by x-radiography, which involves sending x-rays through the material to be inspected and detecting the transmitted x-ray image using a photographic film as shown in Figure G.14. A pore in the material to be inspected is low in density, so it gives more transmitted intensity and greater film darkening. An inclusion of high density in the material gives less transmitted intensity and less film darkening.

Appendix H: Electrochemical Behavior

Electrochemical behavior occurs when there are ions that can move in a medium due to an electric field. The positive ions (cations) move toward the negative end of the voltage gradient, while the negative ions (anions) move toward the positive end (Figure H.1). The medium containing the movable ions is called the "electrolyte." The electrical conductors in contact with the electrolyte for the purpose of applying the electric field are electrodes. The electrode at the positive end of the voltage gradient is the positive electrode or anode. The electrode at the negative end is the negative electrode or cathode. The electrodes must be sufficiently inert so that they do not react with the electrolyte. In order to apply the electric field, the electrodes are connected to a DC power supply such that the cathode is connected to the negative end and the anode to the positive end. In this way, electrons flow in the electrical leads (the outer circuit) from the anode to the cathode, while ions flow in the electrolyte. The electron flow in the outer circuit and the ion flow in the electrolyte constitute a loop of charge flow. Note that the anions flow from cathode to anode in the electrolyte while the electrons flow from anode to cathode in the outer circuit, and both anions and electrons are negatively charged.

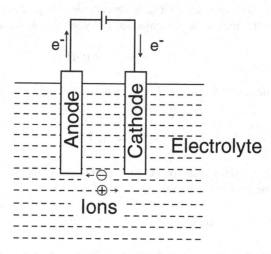

FIGURE H.1 Flow of electrons in the outer circuit from anode to cathode, flow of cations toward cathode, and of anions toward anode in the electrolyte in response to applied voltage, which is positive at the anode.

At the anode, a chemical reaction that gives away one or more electrons takes place. In general, a reaction involving the loss of one or more electrons is known as an "oxidation reaction." The electron lost through this reaction supplies the anode with the electron to release to the outer circuit. An example of an oxidation (anodic) reaction is

$$Cu \rightarrow Cu^{2+} + 2e^- \tag{H.1}$$

where Cu is the anode, which is oxidized to Cu^{2+}, thereby becoming corroded and releasing two electrons to the outer circuit. The Cu^{2+} ions formed remain in the electrolyte, while the electrons go through the anode (which must be electrically conducting) to the outer circuit.

At the cathode, a chemical reaction that takes one or more electrons from the cathode occurs. In general, a reaction involving the gain of one or more electrons is known as a "reduction reaction." The electron gained through this reaction is supplied to the cathode (which must be electrically conducting) by the outer circuit. Always, oxidation occurs at the anode and reduction at the cathode.

The product of the reduction reaction at the cathode may be a species coated on the cathode, as in the electroplating of copper, where the reduction (cathodic) reaction is

$$Cu^{2+} + 2e^- \rightarrow Cu \tag{H.2}$$

where Cu^{2+} is the cation in the electrolyte and Cu is the product of the reduction reaction, and is the species electroplated on the cathode. The product of the reduction reaction may be OH^- ions, as in the case when oxygen and water are present, i.e.,

$$O_2 + 2 H_2O + 4e^- \rightarrow 4 (OH^-) \tag{H.3}$$

The product of the reaction may be H_2 gas, as in the case when H^+ (in an acid) ions are present, i.e.,

$$2H^+ + 2e^- \rightarrow H_2 \uparrow \tag{H.4}$$

The product of the reaction may be H_2 gas together with OH^- ions, as in the case when H_2O is present without O_2, i.e.,

$$2H_2O + 2e^- \rightarrow H_2 \uparrow + 2(OH)^- \tag{H.5}$$

The product of the reaction may be water, as in the case when O_2 and H^+ (an acid) are present, i.e.,

$$O_2 + 4H^+ + 4e^- \rightarrow 2 H_2O \tag{H.6}$$

Which of the abovementioned cathodic reactions takes place depends on the chemical species present. The most common of the reactions is (H.3), as oxygen and water are present in most environments. The variety of cathodic reactions is much greater than that of anodic reactions.

The anodic and cathodic reactions together make up the overall reaction. For example, the anodic reaction (H.1) and cathodic reaction (H.3) together become

$$2 \text{ Cu} \rightarrow 2 \text{ Cu}^{2+} + 4e^-$$
$$O_2 + 2 \text{ H}_2O + 4e^- \rightarrow 4 \text{ (OH}^-)$$

$$\overline{2 \text{ Cu} + O_2 + 2 \text{ H}_2O \rightarrow 2 \text{ Cu}^{2+} + 4(\text{OH}^-)}$$

A Cu^{2+} ion and two OH^- ions formed by the overall reaction react to form $Cu(OH)_2$, which resides in the electrolyte.

Both anode and cathode must be electrically conducting, but they do not have to participate in the anodic or cathodic reactions. The anodic reaction (H.1) involves the anode (Cu) as the reactant, but the cathodic reactions (H.2), (H.3), (H.4), (H.5), and (H.6) do not involve the cathode. Thus, the cathode is usually a platinum wire or other electrical conductor that does not react with the electrolyte.

The electrolyte must allow ionic movement in it. It is usually a liquid, although it can be a solid. For the cathodic reaction (H.2) to occur, the electrolyte must contain Cu^{2+} ions (e.g., a copper sulfate or $CuSO_4$ solution). For the cathodic reaction (H.4) to occur, the electrolyte must contain H^+ ions (e.g., a sulfuric acid or H_2SO_4 solution). The anodic reaction occurs at the interface between the anode and the electrolyte; the cathodic reaction occurs at the interface between the cathode and the electrolyte. The ions that are either reactants (e.g., H^+ in Reaction (H.4)) or reaction products (e.g., OH^- in Reaction (H.3)) must move to or from the electrode/electrolyte interface.

Whether an electrode ends up being the anode or the cathode depends on the relative propensity for oxidation and reduction at each electrode. The electrode at which oxidation has higher propensity becomes the anode and the other becomes the cathode. The propensity depends on the types and concentrations of the chemical species present.

Different species have different propensities to be oxidized. For example, iron ($Fe \rightarrow Fe^{2+} + 2e^-$) has more propensity for oxidation than copper ($Cu \rightarrow Cu^{2+} + 2e^-$), when the Fe^{2+} and Cu^{2+} ions in the electrolyte are at the same concentration. When Fe and Cu are the electrodes and the Fe^{2+} and Cu^{2+} ion concentrations are the same in the electrolyte, Fe will become the anode and Cu will become the cathode (Figure H.2). Fe and Cu constitute a galvanic couple, and this setup constitutes a galvanic cell or composition cell. The anodic reaction will be $Fe \rightarrow Fe^{2+} + 2e^-$, while the cathodic reaction will be (H.2), (H.3), (H.4), (H.5), or (H.6), depending on the relative propensity for these competing reduction reactions.

The concentrations of the species also affect the propensity. In general, high concentrations of the reactants increase the propensity for a reaction, and high concentrations of the reaction products decrease the propensity. This is known as "Le Chatelier's principle." Thus, a high concentration of Cu^{2+} in the electrolyte increases the propensity for Reaction (H.2). When copper is both electrodes, but the Cu^{2+} concentrations are different in the vicinities of the two electrodes, the electrode

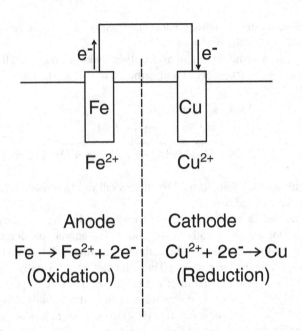

FIGURE H.2 Composition cell with Fe as the anode and Cu as the cathode. Note that electrons and ions flow without the need for an applied voltage between anode and cathode.

with the higher Cu^{2+} ion concentration will become the cathode and the electrode with the lower Cu^{2+} ion concentration will become the anode (Figure H.3). This setup involving a difference in concentration constitutes a concentration cell. Similarly, when copper is both electrodes, but the O_2 concentrations are different in the vicinities of the two electrodes (Figure H.4), the electrode with the higher O_2 concentration will become the cathode and the electrode with the lower O_2 concentration will become the anode. A difference in oxygen concentration is commonly encountered, since oxygen comes from air and access to air can be varied. Therefore, a concentration cell involving a difference in O_2 concentration is particularly common and is called an "oxygen concentration cell." For example, a piece of copper with dirt covering a part of it (Figure H.5) is an oxygen concentration cell, because the part of the copper covered by the dirt has less access to air and becomes the anode, forcing the uncovered part of the copper to be the cathode. Hence, the covered part gets corroded. Similarly, a piece of metal (an electrical conductor) having a crack in it (Figure H.6) is an oxygen concentration cell because the part of the metal at the crack has less access to air and becomes the anode, forcing the part of the metal exposed to open air to be the cathode.

When the two electrodes are open-circuited at the outer circuit, a voltage (called the "open-circuit voltage") exists between them. This voltage describes the difference in the propensity for oxidation between the two electrodes, so it is considered the driving force for the cathodic and anodic reactions that occur when the electrodes are short-circuited at the outer circuit. The greater the difference in oxidation propensity, the higher the open-circuit voltage. This is the same principle behind batteries, which provide a voltage as the output. Because the voltage pertains to the

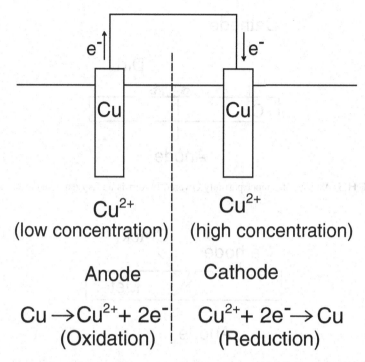

FIGURE H.3 Concentration cell with Cu electrodes, a low Cu^{2+} concentration at the anode side, and a high Cu^{2+} concentration at the cathode side.

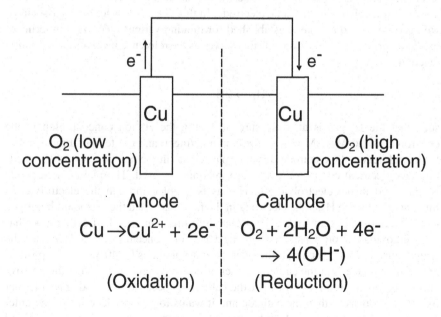

FIGURE H.4 Oxygen concentration cell with Cu electrodes, a low O_2 concentration at the anode side, and a high O_2 concentration at the cathode side.

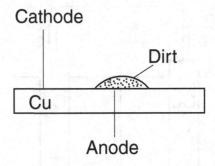

FIGURE H.5 A piece of copper partially covered by dirt is an oxygen concentration cell.

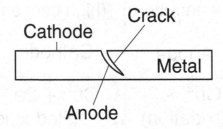

FIGURE H.6 A piece of metal with a crack is an oxygen concentration cell.

difference between a pair of electrodes, it is customary to describe the voltage of any electrode relative to the same electrode. In this way, each electrode is associated with a voltage, and a scale is established for ranking various electrodes in terms of oxidation propensity. The electrode that serves as the reference involves the oxidation reaction

$$H_2 \rightarrow 2H^+ + 2e^-$$

such that the H_2 gas is at a pressure of 1 atm, the H^+ ion concentration in the electrolyte is 1.0 M (1 M = 1 molar = 1 mole/liter), and the temperature is 25°C. In this reference, the electrode does not participate in the reaction, so it only needs to be an electrical conductor. Therefore, platinum is used. The platinum electrode is immersed in the electrolyte and H_2 gas is bubbled through the electrolyte, as illustrated in Figure H.7. The electrode in the figure is called the "standard hydrogen reference half-cell" (half of a cell). Relative to this reference half-cell, copper has a lower oxidation propensity (when the Cu^{2+} ion concentration is 1 M and the temperature is 25°C) such that the open circuit voltage is 0.340 V and the positive end of the voltage is at the cathode, which is copper (Figure H.8). That the positive end of the open circuit voltage is at the cathode is because the anode has a higher oxidation propensity than the cathode and it wants to release electrons to the outer circuit, which cannot accept them because of the open circuit situation. The standard electrode potential for copper is thus +0.340 V. A ranking of the electrodes in terms

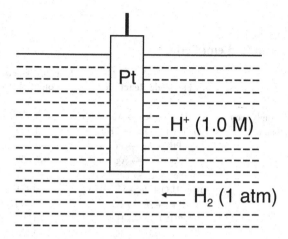

FIGURE H.7 The standard hydrogen reference half-cell.

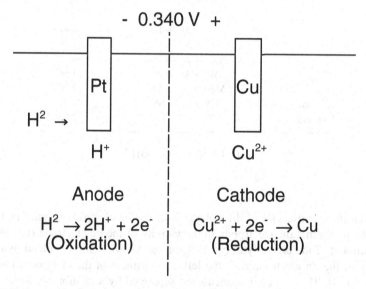

FIGURE H.8 A cell with copper as the cathode and the hydrogen half-cell as the anode under open-circuit condition. The open-circuit voltage is 0.340 V — positive at the cathode.

of the standard electrode potential is shown in Table H.1, which is known as the "standard electromotive force (emf) series." Note that this ranking is for the case in which the ion concentration is 1 M and the temperature is 25°C. If the ion concentration or the temperature is changed, the ranking can be different. The higher the electrode is in Table H.1, the more cathodic it is. The lower the electrode is in Table H.1, the more anodic it is. For example, from Table H.1, Fe is cathodic relative to Zn, and the open-circuit voltage between the Fe and Zn electrodes (called the "overall cell potential") is

TABLE H.1
The Standard emf Series

	Electrode Reaction	Standard Electrode Potential (V)
Increasingly inert	$Au^{3+} + 3e^- \rightarrow Au$	+1.420
(Cathodic)	$O_2 + 4H^+ + 4e^- \rightarrow 2H_2O$	+1.229
	$Pt^{2+} + 2e^- \rightarrow Pt$	~+1.2
	$Ag^+ + e^- \rightarrow Ag$	+0.800
	$Fe^{3+} + e^- \rightarrow Fe^{2+}$	+0.771
	$O_2 + 2H_2O + 4e^- \rightarrow 4(OH^-)$	+0.401
	$Cu^{2+} + 2e^- \rightarrow Cu$	+0.340
	$2H^+ + 2e^- \rightarrow H_2$	0.000
	$Pb^{2+} + 2e^- \rightarrow Pb$	−0.126
	$Sn^{2+} + 2e^- \rightarrow Sn$	−0.136
	$Ni^{2+} + 2e^- \rightarrow Ni$	−0.250
	$Co^{2+} + 2e^- \rightarrow Co$	−0.277
	$Cd^{2+} + 2e^- \rightarrow Cd$	−0.403
	$Fe^{2+} + 2e^- \rightarrow Fe$	−0.440
	$Cr^{3+} + 3e^- \rightarrow Cr$	−0.744
	$Zn^{2+} + 2e^- \rightarrow Zn$	−0.763
	$Al^{3+} + 3e^- \rightarrow Al$	−1.662
	$Mg^{2+} + 2e^- \rightarrow Mg$	−2.363
Increasingly active	$Na^+ + e^- \rightarrow Na$	−2.714
(Anodic)	$K^+ + e^- \rightarrow K$	−2.924

$$[(-0.440) - (-0.763)] \text{ V}$$

$$= 0.323 \text{ V}$$

such that the positive end of the voltage is at the Fe electrode (Figure H.9). Note from the figure that the electrolyte in the vicinity of the Fe electrode (i.e., the right compartment of the electrolyte) has Fe^{2+} ions at 1.0 M, while the electrolyte in the vicinity of the Zn electrode (i.e., the left compartment of the electrolyte) has Zn^{2+} ions at 1.0 M. The two compartments are separated by a membrane, which allows ions to flow through (otherwise it would be open circuited within the electrolyte) but provides enough hindrance to the mixing of the electrolytes in the two compartments. The anode reaction in Figure H.9 is

$$Zn \rightarrow Zn^{2+} + 2e^-$$

The cathode reaction is

$$Fe^{2+} + 2e^- \rightarrow Fe$$

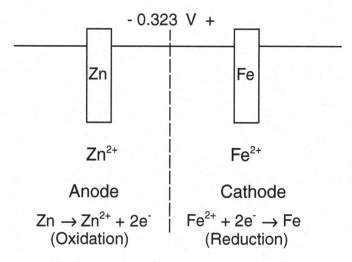

FIGURE H.9 A cell with zinc as the anode and iron as the cathode under open-circuit condition. The open-circuit voltage is 0.323 V — positive at the cathode.

The overall reaction is

$$Zn + Fe^{2+} \rightarrow Zn^{2+} + Fe$$

The cell of Figure H.9 is commonly written as

$$Zn \,|\, Zn^{2+} \,|\, Fe^{2+} \,|\, Fe \qquad\qquad (H.7)$$

where Zn and Fe^{2+} are the reactants, Zn^{2+} and Fe are the reaction products, and the vertical lines denote phase boundaries.

Appendix I:
The pn Junction

A pn junction functions as a diode, with the current-voltage characteristic shown in Figure I.1. The pn junction allows current to go from the p-side to the n-side (I > 0, V > 0, forward bias), and almost no current from the n-side to the p-side (I < 0 but almost zero when V < 0, reverse bias). When the applied voltage V is very negative, a large negative current flows. This is known as "breakdown."

Consider a pn junction at V = 0 (open circuited). Because the hole concentration is much higher in the p-side than the n-side, holes diffuse from the p-side to the n-side, causing the exposure of some acceptor anions near the junction in the p-side (Figure I.2). Similarly, because the conduction electron concentration is much higher in the n-side than the p-side, the conduction electrons diffuse from the n-side to the p-side, thus causing the exposure of some donor cations near the junction in the n-side. In this way, diffusion results in a region with very few carriers near the junction. This region is the depletion region (or the space-charge layer, or the dipole layer, or the transition region). The exposed ions in the depletion region cause an electric field such that the electric potential is higher in the n-side than the p-side. The difference in electric potential between the two sides is called the "contact potential" (V_o). Note that the contact potential is present even when the applied voltage V is zero. The contact potential is a barrier for the diffusion of holes from the p-side to the n-side because holes want to go down in potential. Similarly, the contact potential is a barrier for the diffusion of conduction electrons from the n-side to the p-side because electrons want to go up in potential. Both the diffusion of holes from the p-side to the n-side and the diffusion of electrons from the n-side to the p-side contribute to the diffusion current I_d, which flows from the p-side to the n-side.

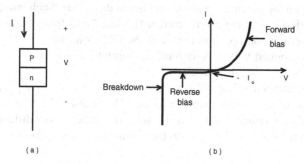

(a)　　　　　　　　　　(b)

FIGURE I.1 The pn junction (**a**) configuration, (**b**) current (I)-voltage (V) relationship.

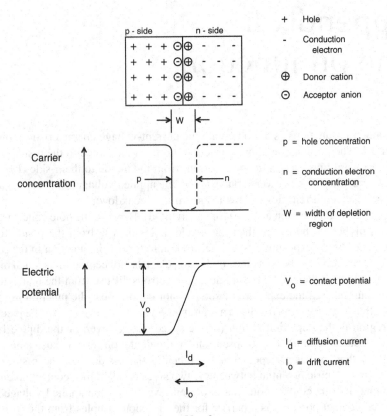

FIGURE I.2 A pn junction at bias voltage V = 0.

There is a small concentration of holes in the n-side. When these holes approach the depletion region, they spontaneously go down the potential gradient and get to the p-side. Similarly, there is a small concentration of conduction electrons in the p-side. When these conduction electrons approach the depletion region, they spontaneously go up the potential gradient and get to the n-side. Such movements of the minority carriers constitute a drift current (I_o) that flows from the n-side to the p-side, i.e., in a direction opposite to that of the diffusion current I_d.

When the applied voltage V is zero, the current I is also zero, since there is an open circuit. Under this situation, $I_d = I_o$.

When the applied voltage V is negative (reverse bias), the applied potential is more positive in the n-side than the p-side. This causes the contact potential to increase to $V_o - V$ (since V < 0). As a result, the barrier to the diffusion current is increased, thus greatly lowering I_d, to the extent that $I_d \ll I_o$. Hence,

$$I = I_d - I_o \cong -I_o$$

Because I_o is due to the minority carriers, it is bound to be small. Therefore, under reverse bias, a very small current flows from the n-side to the p-side.

When the applied voltage V is very negative, the contact potential becomes very large, resulting in an intense electric field in the depletion region. This strong electric field may tear electrons out of the covalent bonds, creating many more holes and conduction electrons. The holes and electrons get swept across the depletion region with high kinetic energies, knocking more valence electrons out of covalent bonds. The consequence is a large reverse current ($I < 0$, $V < 0$). This phenomenon is known as "breakdown."

When the applied voltage V is positive (forward bias), the applied potential is more positive in the p-side than the n-side, so the contact potential decreases to $V_o - V$. Hence, the barrier to the diffusion current is lowered and $I_d \gg I_o$. Therefore,

$$I = I_d - I_o \approx I_d$$

In fact, I_d increases exponentially with increasing V.

Appendix J: Carbon Fibers

Carbon fibers are fibers that are at least 92 wt.% carbon in composition.[1] They can be short or continuous; their structure can be crystalline, amorphous, or partly crystalline. The crystalline form has the crystal structure of graphite (Figure J.1), which consists of sp^2 hybridized carbon atoms arranged two-dimensionally in a honeycomb structure in the x-y plane. Carbon atoms within a layer are bonded by (1) covalent bonds provided by the overlap of the sp^2 hybridized orbitals, and (2) metallic bonding provided by the delocalization of the p_z orbitals, i.e., the π electrons. This delocalization makes graphite a good electrical conductor and a good thermal conductor in the x-y plane. The bonding between the layers is van der Waals bonding, so the carbon layers can easily slide with respect to one another; graphite is an electrical insulator and a thermal insulator perpendicular to the layers. Due to the difference between the in-plane and out-of-plane bonding, graphite has a high modulus of elasticity parallel to the plane and a low modulus perpendicular to the plane. Thus, graphite is highly anisotropic.

The high modulus of a carbon fiber stems from the fact that the carbon layers, though not necessarily flat, tend to be parallel to the fiber axis. This crystallographic preferred orientation is known as a "fiber texture." A carbon fiber has a higher modulus parallel to the fiber axis than perpendicular to the fiber axis, and the coefficient of thermal expansion is lower along the fiber axis.

The greater the degree of alignment of the carbon layers parallel to the fiber axis (i.e., the stronger the fiber texture, the greater the c-axis crystallite size (L_c), the density, the carbon content, and the fiber's tensile modulus, electrical conductivity, and thermal conductivity parallel to the fiber axis), the smaller the fiber's coefficient of thermal expansion and internal shear strength.

The carbon layers in graphite are stacked in an AB sequence such that half of the carbon atoms have atoms directly above and below them in adjacent layers (Figure J.1). Note that this AB sequence differs from that in a hexagonal close-packed (HCP) crystal structure. In a carbon fiber, there can be graphite regions of size L_c perpendicular to the layers and size L_a parallel to the layers. There can also be crystalline regions in which the carbon layers, though well developed and parallel to one another, are not stacked in any particular sequence; the carbon in these regions is said to be turbostratic carbon. Yet another type of carbon that can exist in carbon fibers in amorphous carbon, in which the carbon layers, though well developed, are not even parallel to one another.

The proportion of graphite in a carbon fiber can range from 0 to 100%. When the proportion is high, the fiber is said to be graphitic, and it is called a graphite

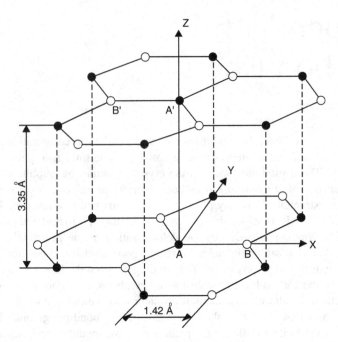

FIGURE J.1 The crystal structure of graphite.

fiber. However, a graphite fiber is polycrystalline, whereas a graphite whisker is a single crystal with the carbon layer rolled up like a scroll. Because of their single crystal nature, graphite whiskers are virtually flaw-free and have exceptionally high strength. However, the production yield of graphite whiskers is too low for them to be commercially significant.

Table J.1[2] compares the mechanical properties, melting temperature, and density of carbon fibers with other types of fibers. There are numerous grades of carbon fibers; Table J.1 only shows the two high-performance grades, which are labeled "high strength" and "high modulus." Among the fibers (not counting the whiskers), high-strength carbon fibers exhibit the highest strength, while high-modulus carbon fibers exhibit the highest elasticity. Moreover, the density of carbon fibers is quite low, making the specific modulus (modulus/density ratio) of high-modulus carbon fibers exceptionally high. The polymer fibers, such as polyethylene and Kevlar® fibers, have densities even lower than carbon fibers, but their melting temperatures are low. The ceramic fibers, such as glass, SiO_2, Al_2O_3, and SiC fiber, have densities higher than carbon fibers; most of them (except glass fibers) suffer from high prices or are not readily available in a continuous fiber form. The tensile stress-strain curves of the fibers are straight lines all the way to fracture, so the strength divided by the modulus gives the ductility (strain at break) of each fiber, as shown in Table J.1. The main drawback of the mechanical properties of carbon fibers is in the low ductility, which is lower than those of glass, SiO_2, and Kevlar fibers. The ductility of high-modulus carbon fibers is even lower than that of high-strength carbon fibers.

TABLE J.1
Properties of Various Fibers and Whiskers

Material	Density[a] (g/cm³)	Tensile Strength[a] (GPa)	Modulus of Elasticity[a] (GPa)	Ductility (%)	Melting Temp.[a] (°C)	Specific Modulus[a] (10⁶ m)	Specific Strength[a] (10⁴ m)
E-glass	2.55	3.4	72.4	4.7	< 1725	2.90	14
S-glass	2.50	4.5	86.9	5.2	< 1725	3.56	18
SiO₂	2.19	5.9	72.4	8.1	1728	3.38	27.4
Al₂O₃	3.95	2.1	380	0.55	2015	9.86	5.3
ZrO₂	4.84	2.1	340	0.62	2677	7.26	4.3
Carbon (high-strength)	1.50	5.7	280	2.0	3700	18.8	19
Carbon (high-modulus)	1.50	1.9	530	0.36	3700	36.3	13
BN	1.90	1.4	90	1.6	2730	4.78	7.4
Boron	2.36	3.4	380	0.89	2030	16.4	12
B₄C	2.36	2.3	480	0.48	2450	20.9	9.9
SiC	4.09	2.1	480	0.44	2700	12.0	5.1
TiB₂	4.48	0.10	510	0.02	2980	11.6	0.3
Be	1.83	1.28	300	0.4	1277	19.7	7.1

TABLE J.1 (continued)
Properties of Various Fibers and Whiskers

Material	Density[a] (g/cm³)	Tensile Strength[a] (GPa)	Modulus of Elasticity[a] (GPa)	Ductility (%)	Melting Temp.[a] (°C)	Specific Modulus[a] (10⁶ m)	Specific Strength[a] (10⁴ m)
W	19.4	4.0	410	0.98	3410	2.2	2
Polyethylene	0.97	2.59	120	2.2	147	12.4	27.4
Kevlar®	1.44	4.5	120	3.8	500	8.81	25.7
Al_2O_3 whiskers	3.96	21	430	4.9	1982	11.0	53.3
BeO whiskers	2.85	13	340	3.8	2550	12.3	47.0
B_4C whiskers	2.52	14	480	2.9	2450	19.5	56.1
SiC whiskers	3.18	21	480	4.4	2700	15.4	66.5
Si_3N_4 whiskers	3.18	14	380	3.7	—	12.1	44.4
Graphite whiskers	1.66	21	703	3.0	3700	43	128
Cr whiskers	7.2	8.90	240	3.7	1890	3.40	12

[a] From Ref. 2.

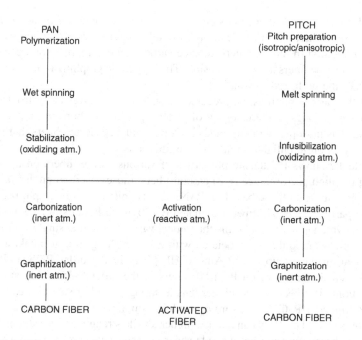

SCHEME J.1 The process for making carbon fibers from PAN and pitch precursors.

Carbon fibers that are commercially available are divided into three categories, namely general-purpose (GP), high-performance (HP), and activated carbon fibers (ACF). The general-purpose type is characterized by an amorphous and isotropic structure, low tensile strength, low tensile modulus, and low cost. The high-performance type is characterized by relatively high strength and modulus. Among the high-performance carbon fibers, a higher modulus is associated with a higher proportion of graphite and more anisotropy. Activated carbon fibers are characterized by the presence of a large number of open micropores, which act as adsorption sites. The adsorption capacity of activated carbon fibers is comparable to that of activated carbons, but the fiber shape of activated carbon fibers allows the adsorbate to get to the adsorption site faster, thus accelerating the adsorption and desorption processes.[3] The amount adsorbed increases with the severity of activation. Severe activation may be achieved by treating commercial ACF with sulfuric acid, followed by heating at up to 500°C.[4]

Commercial carbon fibers are fabricated by using pitch or polyacrylonitrile (PAN) as the precursor. The process for both precursors is shown in Scheme J.1.[5]

Precursor fibers are fabricated by conventional spinning techniques, such as wet spinning for PAN and melt spinning for pitch. They must be converted to a form that is flameproof and stable at the high temperatures (> 700°C) involved in carbonization. Therefore, before carbonization (pyrolysis), they are stabilized in the case of the PAN precursor, or infusiblized in the case of the pitch precursor. Both stabilization and infusiblization are carried out in an oxidizing atmosphere. After that, general-purpose and high-performance fibers are obtained by carbonization in an inert atmosphere, followed by graphitization at > 2500°C in an inert atmosphere if a high modulus is desired. Activated carbon fibers are obtained by activation in a reactive atmosphere, such as steam at elevated temperatures. To enhance the preferred orientation in the high-performance carbon fibers, graphitization can be performed while the fibers are under tension. The higher the graphitization temperature, the greater the preferred orientation.

For the case of pitch as the precursor, isotropic pitch gives an isotropic carbon fiber, which belongs to the category of general-purpose carbon fibers. Anisotropic pitch (such as mesophase pitch) yields high-performance carbon fibers that have the carbon layers preferentially parallel to the fiber axis.

Table J.2 shows the tensile properties of various carbon fibers on the market. Among the high-performance (HP) carbon fibers, those based on pitch can attain a higher modulus than those based on PAN, because pitch is more graphitizable than PAN. In particular, the HP fiber designated E-130 by du Pont exhibits a modulus of 894 GPa, which is over 80% of the theoretical value of graphite single crystal (1000 GPa). A higher modulus is associated with a lower elongation at break, as shown by comparing the group of BP Amoco HP fibers in the order P-25, P-75S, and P-120S, and the group of du Pont HP fibers in the order E-35, E-75, and E-130. The du Pont HP fibers exhibit higher tensile strengths and greater elongations than the BP Amoco HP fibers of similar moduli. Among the high-performance (HP) fibers, those based on PAN can attain a higher tensile strength and greater elongation than those based on pitch because (1) shear is easier between the carbon layers in a graphitized fiber, (2) pitch is more graphitizable than PAN, and (3) the oriented graphitic structure causes the fibers to be more sensitive to surface defects and structural flaws. In particular, the HP PAN-based fiber designated T-1000 by Toray exhibits a tensile strength of 7060 MPa and an elongation of 2.4%. The general-purpose (GP) fibers tend to be low in strength and modulus, but high in elongation at break.

Table J.2 also shows the diameters of various commercial carbon fibers. Among the HP fibers, those based on PAN have smaller diameters than those based on pitch.

Pitch-based carbon fibers (GP and HP) represent only about 10% of the total carbon fibers produced worldwide around 1990,[6] but this percentage is increasing due to the lower cost and higher carbon content of pitch compared to PAN. The costs of precursors and carbon fibers are shown in Table J.3. Mesophase pitch-based carbon fibers are currently the most expensive because of the processing cost. Isotropic pitch-based carbon fibers are the least expensive. PAN-based carbon fibers are intermediate in cost.

Although carbon fibers are mostly more expensive than aramid fibers or glass fibers, they provide higher tensile strengths. Among the different grades of carbon fibers, the prices differ greatly. In general, the greater the tensile strength, the higher the price.[7]

The price of carbon fibers has been decreasing while consumption has been increasing.[7] The decreasing price is broadening the applications of carbon fibers from military to civil application, from aerospace to automobile applications, and from biomedical devices to concrete structures.

TABLE J.2
Tensile Properties and Diameters of Commercial Carbon Fibers

Type	Fiber Designation	Tensile Strength (MPa)	Tensile Modulus of Elasticity (GPa)	Elongation at Break (%)	Diameter (μm)	Manufacturer
GP	T-101S	720	32	2.2	14.5	Kureha Chem.
	T-201S	690	30	2.1	14.5	Kureha Chem.
	S-210	784	39	2.0	13	Donac
	P-400	690	48	1.4	10	Ashland Petroleum
	GF-20	980	98	1.0	7–11	Nippon Carbon
HP (PAN)	T-300	3530	230	1.5	7.0	Toray
	T-400H	4410	250	1.8	7.0	Toray
	T-800H	5590	294	1.9	5.2	Toray
	T-1000	7060	294	2.4	5.3	Toray
	MR 50	5490	294	1.9	5	Mitsubishi Rayon
	MRE 50	5490	323	1.7	6	Mitsubishi Rayon
	HMS-40	3430	392	0.87	6.2	Toho Rayon
	HMS-40X	4700	392	1.20	4.7	Toho Rayon
	HMS-60X	3820	588	0.65	4.0	Toho Rayon
	AS-1	3105	228	1.32	8	Hercules
	AS-2	2760	228	1.2	8	Hercules
	AS-4	3795	235	1.53	8	Hercules
	AS-6	4140	242	1.65	5	Hercules
	IM-6	4382	276	1.50	5	Hercules

TABLE J.2 (CONTINUED)
Tensile Properties and Diameters of Commercial Carbon Fibers

Type	Fiber Designation	Tensile Strength (MPa)	Tensile Modulus of Elasticity (GPa)	Elongation at Break (%)	Diameter (μm)	Manufacturer
	HMS4	2484	338	0.7	8	Hercules
	HMU	2760	380	0.70	8	Hercules
HP (pitch)	P-25	1400	160	0.9	11	BP Amoco
	P-75S	2100	520	0.4	10	BP Amoco
	P-120S	2200	827	0.27	10	BP Amoco
	E-35	2800	241	1.03	9.6	du Pont
	E-75	3100	516	0.56	9.4	du Pont
	E-130	3900	894	0.55	9.2	du Pont
	F-140	1800	140	1.3	10	Donac
	F-600	3000	600	0.52	9	Donac
ACF	FX-100	—	500[a]	18[b]	15	Toho Rayon
	FX-600	—	1500[a]	50[b]	7	Toho Rayon
	A-10	245	1000[a]	20[c]	14	Donac
	A-20	98	2000[a]	45[c]	11	Donac

[a] Specific surface area (m²/g)
[b] Adsorption amount of benzene (%)
[c] Adsorption amount of acetone (%)

TABLE J.3
Cost of PAN-Based, Mesophase Pitch-Based, and Isotropic Pitch-Based Carbon Fibers (From Ref. 6.)

	Cost of Precursor ($/kg)	Cost of Carbon Fibers ($/kg)
PAN-based	0.40	60
Mesophase pitch-based	0.25	90
Isotropic pitch-based	0.25	22

Under rapid development are short carbon fibers grown from the vapor of low-molecular-weight hydrocarbon compounds, such as acetylene. This process involves catalytic growth using solid catalyst particles (e.g., Fe) to form carbon filaments, which can be as small as 0.1 μm in diameter. Subsequent chemical vapor deposition from the carbonaceous gas in the same chamber causes the filaments to grow in diameter, resulting in vapor-grown carbon fibers (VGCF) or gas-phase-grown carbon fibers.

Carbon fibers can alternatively be classified on the bases of their tensile strength and modulus. The nomenclature given below was formulated by IUPAC.

- UHM (ultra high modulus) type: carbon fibers with modulus greater than 500 GPa
- HM (high modulus) type: carbon fibers with modulus greater than 300 GPa and strength-to-modulus ratio less than 1%
- IM (intermediate modulus) type: carbon fibers with modulus up to 300 GPa and strength-to-modulus ratio above 1×10^{-2}
- Low-modulus type: carbon fibers with modulus as low as 100 GPa and low strength; they have an isotropic structure
- HT (high strength) type: carbon fibers with strength greater than 3 GPa and strength-to-modulus ratio between 1.5 and 2×10^{-2}

There is overlap between the IM and HT categories, as shown by the above definitions.

Commercial continuous carbon fibers are in the form of tows (untwisted bundles) containing typically 1,000–12,000 fibers (filaments) per tow, or yarns (twisted bundles). They may be sized or unsized. The sizing improves the handleability and may enhance the bonding between the fibers and certain matrices when the fibers are used in composites.

High-performance carbon fibers are widely used in polymer-matrix composites for aircraft that are lightweight for the purpose of saving fuel. The aircraft Voyager, which has 90% of its structure made of such composites, achieved a nonstop, unfueled, around-the-world flight in 1986. The use of such composites in passenger aircraft is rapidly increasing. High-performance carbon fibers are also used in carbon-matrix composites for high-temperature aerospace applications, such as the Space Shuttle, as the carbon matrix is more temperature resistant than a polymer matrix. These fibers have begun to be used in metal matrices, such as aluminum, for aerospace applications, as aluminum is more temperature resistant than polymers.

Short general-purpose pitch-based carbon fibers are used for the reinforcement of concrete, because low cost is crucial for the concrete industry. Because this is a large-volume application of carbon fibers, the tonnage of carbon fibers used is expected to increase markedly as this application becomes more widely accepted. General-purpose carbon fibers are also used for thermal insulation, sealing materials, electrically conducting materials, antistatic materials, heating elements, electrodes, filters, friction materials, sorbents, and catalysts.[8]

REFERENCES

1. E. Fitzer, in *Carbon Fibers, Filaments, and Composites*, J.L. Figueiredo, C.A. Bernardo, R.T.K. Baker, and K.J. Huttinger, Eds., Kluwer Academic, Dordrecht, pp. 3-41 (1990).
2. D.R. Askeland, *The Science and Engineering of Materials*, 2nd Ed., PWS-Kent, Boston, p. 591 (1989).
3. L.I. Fridman and S.F. Grebennikov, *Khimicheskie Volokna* 6, 10-13 (1990).
4. Isao Mochida and Shizuo Kawano, *Ind. Eng. Chem. Res.* 30(10), 2322-2327 (1991).
5. K. Okuda, *Trans. Mater. Res. Soc. Jpn.* 1, 119-139 (1990).
6. D.D. Edie, in *Carbon Fibers, Filaments, and Composites*, J.L. Figueiredo, C.A. Bernardo, R.T.K. Baker, and K.J. Huttinger, Eds., Kluwer Academic, Dordrecht, pp. 43-72 (1990).
7. E. Fitzer and F. Kunkele, *High Temp. High Pressures* 22(3), 239-266 (1990).
8. R.M. Levit, *Khimicheskie Volokna* 6, 16-18 (1990).

Index

Printed in the United States
by Baker & Taylor Publisher Services